MAURIC.
SAXE'
1745 CAMP
IN BELG1

By Henry Pichɛ
Translated by G. F. N

Original work:

PARIS

Libraire Militaire R. Chapei
30 Rue et Passage Dauphine

. . . .

1909

Maurice De Saxe's 1745 Campaign in Belgium By Henry Pichat
Translated by George Nafziger
Cover of Maurice De Saxe
This edition published in 2022

Winged Hussar is an imprint of

Winged Hussar Publishing, LLC
1525 Hulse Rd, Unit 1
Point Pleasant, NJ 08742

Copyright © Winged Hussar Publishing
ISBN 978-1-950423-99-6 HC
ISBN 978-1-958872-08-6 EB
LCN 2022947832

Bibliographical References and Index
1. History. 2. Maurice. 3. Military

Winged Hussar Publishing, LLC All rights reserved
For more information
visit us at www.whpsupplyroom.com

Twitter: WingHusPublLC
Facebook: Winged Hussar Publishing LLC

Table of Contents

INTRODUCTION

Since 1741, the date of the beginning of the War of the Austrian Succession, until 1744, the battlefields of Bavaria and Bohemia had, almost alone, held the attention of Europe. In 1745, events would give a great importance to two new theaters: Flanders and Germany.

The first opened on the northern French frontier which was more or less free at the end of 1744, because the French armies, in effect, had carried Menin, Ypres, and Furnes from Maria Theresa, Queen of Hungary. These conquests had broken the barrier of fortresses that the Treaty of Utrecht had raised to contain the ambition of the successors of Louis XIV. In April 1745, French troops assembled below Maubeuge prepared to continue the work begun and to besiege Tournai.

This campaign was barely opened when the brilliant victory at Fontenoy, followed by the surrender of Tournai, quickly enlarged the breach opened in 1744. The Dutch saw a new hole appearing in the wall that covered their frontier and Maria Theresa lost another bit of the inheritance that her forefathers had bequeathed to her in the Lowlands.

In London, our cannon had cruelly resounded. They had cut bloody holes in the British army it fought, on 11 May, at Fontenoy, with such stubbornness. They had also marked the collapse of the old military reputation that the English had acquired on the continent since Crécy. These same Lowlands, still filled with the memories of the exploits of William of Orange and Marlborough, now saw appear the new glories of Louis XV and Maurice de Saxe. Thirty years had barely passed since France had suffered two great defeats. Ramillies had tarnished the brilliance of French arms, Malplaquet had opened the French frontier to invasion. Fontenoy re-established all the prestige of the French Army and opened to it the Lowlands. The King now moved, no doubt, to follow up on the so happily begun campaign in Flanders.

Other events would give the campaign in Germany an importance no less considerable.

Henry Pichat

The Emperor Charles VII, seated on the throne of the King of the Romans since January 1742, died on 20 February 1745. The Diet gathered and proceeded to hold a new Election. Maria Theresa had resolved to assure sufficient votes for her spouse François de Lorraine. Without delay, 30,000 Dutch and Austrians were assembled before the French troops of Maréchal Maillebois camped on the Main. Louis XV, in effect, had become the defender of the rights and liberties of the electoral corps, whose independence was now threatened by the arms and diplomacy of Maria Theresa. The King of France was opposed to the idea that the Vienna Court could dispose of, at its pleasure, the Imperial Crown.

His Most Christian Majesty had two important Allies. The first, Frederick II, King of Prussia, was master of Silesia, but the Austrian army of Charles de Lorraine attempted to take it back from him. The conservation of his precious conquest constituted the principal occupation of the Prussian King. He saw, with much ill humor, Louis XV put himself at the head of the Army of Flanders. There, where the King paid with his person, there also were united the best and most numerous troops. Frederick could never conceal that he preferred to see the French "pushed into the oven" in Germany, according to his own expression of his desire. He translated his discontent into a sort of ultimatum that appeared at Versailles at the beginning of May 1745, and he saw, in the victory at Fontenoy, a definitive response to his complaints. So during this war the French had to, as one said later, work for "the King of Prussia", but it would not be in Germany, during 1745, at least. It was necessary to fear, as a result, a new defection in the Prussian alliance.

France's second ally, Maximilian of Bavaria, son of the dead Emperor, was actively solicited by Maria Theresa's diplomats. The Queen of Hungary demanded that the young Prince renounce any secular pretensions of his family for the Imperial Throne and that he give his vote to François de Lorraine. The Austrian Bathyanyi camped at the gates of Bavaria. He did not wait, to force them as a gesture to his sovereign. With his 25,000 soldiers, he pressed the indecisive Maximilian, who was divided between the promises of Louis XV and the threats of the Queen; he was torn between ambition and fear. What would be the attitude of the French Army of Germany in the middle of all these difficulties?

Maurice De Saxe's 1745 Campaign in Belgium

It does not enter within the framework of this study to follow the discussion of the argument of the "d'Argenson of the war", no more than those of the "d'Argenson of peace", to decide if Louis XV, resolved to carry out an "energetic war" in Flanders, would not have been better inspired by executing a vigorous offensive in Germany. We only propose to follow the military operations carried out in the Netherlands by the last of the great captains who led the armies of the old monarchy.

Notes on this edition and its translation:

The original work is entitled *Campagne de maréchal de Saxe Dans les Flandres: De Fontenoy (Mai 1745) à la prise de Bruxelles (Fevrier 1746) [Campaign of Marshal de Saxe in the Flanders: From Fontenoy (May 1745) to the Capture of Brussels (February 1746)].*

The document I worked with had all the maps described in the text. In examining the supporting documents included with this work, it was decided that they did not add significantly to the interest that the modern reader might have in this work; that they would be of more use to the historian, who would be better served to read them in the original French. However, all orders of battle, states, and returns have been retained, as they are of interest to the modern history aficionado.

This work was written from the French perspective, so the words "we" and "our" were repeatedly encountered. The work was re-written so that these phrases reference "the French", etc., as appropriate. In addition, the term "enemy" has been changed, where possible, to the proper nationality or the more generic "Allies." The structure of these sentences was, again, revised, except in quotations, to permit this change.

Henry Pichat

CHAPTER I
The Allied Army from 11 May to the end of June 1745.

Austrian Netherlands 1745

On 11 May 1745, the action began before Fontenoy at 5:00 a.m., and continued as is well known. Towards 1:00 p.m., at the moment when the Prince von Waldeck, commanding the Allies' left wing, prepared to attempt a new effort, a brief note, signed Guillaume, stopped him. "My Prince," he wrote to the Duke of Cumberland, "I am retiring under the cannons of Ath."[1]

The battle was, in effect, lost. The lack of success of the attacks directed against the redoubt and the Fontenoy cemetery, the definitive and bloody check of the legendary Britannic column crushed by French artillery, pierced by the furious charges of the Maison du Roi and the Carabiniers, obliged the Anglo-Hanoverians to abandon

---

[1]Waldeck, 11 May.

the battlefield. [2]

Waldeck pushed his troops into their camp. He thought that a more complete retreat was not urgent. "I was with the Duke," he said, "to dissuade him from marching that evening to Ath and to resume this march the following day, since it did not appear that the enemy made any dispositions to pursue us, and that a night march could fall into confusion." These representations had no effect. From 7:00 pm., the Duke began his movement and at 9:00 p.m., the Dutch had to leave in their turn. They passed the night on the Catoire stream, near Leuze, and on the 11th, at 8:00 a.m., they reached Ath, which the English had gained the day before "in truth." [3]

On the 16th, after some preparations, the complete Pragmatic Army extended from Lessines to Rebaix, on the left bank of the Dender. [4] In reality, it supported itself on Ath, because an important detachment of cavalry camped on the weak heights that ran along the river between that place and Rebaix. [5] Thus the Allies occupied a front of a dozen kilometers. Their right connected with Audenarde by five leagues of a mediocre road crossing Grammont, a small fortress situated on a strong position at the swampy confluence of the Dender and the Marcq. Some English troops occupied Schoorisse and Everbecq to cover these important communications.

Established at Chièvres and Herchies, Dutch parties assured the liaison of the Allied left with Mons, along a good road of 20 kilometers. [6] Constructed like a granary in the middle of the rich and fertile Austrian Hainaut, this fortress contained considerable provisions. Its strategic situation further augmented its importance. Provided with excellent defenses, it barred the Haine Valley, overlooking the French fortresses of Maubeuge and Condé, and connected directly to Brussels, to Charleroi, and the Valley of the Sambre by Binch, assuring the communications of the Pragmatic Army with Holland and Germany.

Some light troops left at Leuze and Ellegnies watched the debouches from Tournai. [7] Others, established defensively in the de Ligne and the Hainaide Châteaus, patrolled daily in the dense woods of Saint-Amand, covering the front of the Pragmatic Army. [8]

[2] Waldeck, 11 May.
[3] Waldeck, 12 May.
[4] Waldeck, 16 May.
[5] Waldeck, 17 May.
[6] Waldeck, 19-20 May.
[7] Waldeck, 23 May.
[8] Waldeck, 21 May.

In this position, the Allies endeavored, to the end of June, to repair the losses that they had so unexpectedly sustained at the beginning of the campaign.

Before the battle, their army, in the field, contained 41,250 combatants.[9] It lost, on 11 May, 7,199 men killed, wounded, or disappeared, to which one must add the approximately 1,000 prisoners lost during the retreat.[10] Thus reduced, could it act effectively on a defensive front running from the Sambre River to the sea, across a flat land with little cover, sprinkled, in truth, with some strong places, like Audenarde, Ath, Mons, and Namur, but containing no natural defensive position beyond the course of the Dender?

"I must show again to Your Protestant Highness, that it is of the greatest importance and necessity to reinforce this army and beg them to send us, as soon as possible, the seven battalions that they have held with such prudence in reserve," wrote Marshal von Königsegge, to La Haye, on the day after the defeat.[11]

After 20 May, new Dutch troops arrived, followed, during the course of June, by other contingents coming from Wenloo, Bergen-op-Zoom, Nijmegen, Arnhem, Maestricht, and Hertogenbosch.[12] The Allies also drew them from the fortresses on the defensive front, Namur, Charleroi, and Mons.[13]

Nothing escaped the vigilance of the French agents. The Allies had barely installed themselves in their camp at Lessines that Minister d'Argenson took dispositions to be informed on all their movements.[14] Commandants of the frontier fortresses, Intendants of the provinc-

---

[9] Order of battle of the Allied army at the battle of Fontenoy. – Another document conserved in the Corresp. A.F: Order of battle of the Allied Army in Flanders, 11 May, gives figures that are slightly different.

[10] List of dead, wounded, or disappeared by the Allied army in the attack on the French camp near Fontenoy (Waldeck). – List of the Allies killed or taken prisoner during the retreat (Corresp. A.F.).

[11] Königsegg to LL. HH. PP., Lessines, 15 May (Corresp. G.H.).

[12] Waldeck, May-June.

[13] Waldeck, May-June.

[14] "Informed of all the movements of the troops which you know; so that the letters of the intermediaries are not delayed, have them passed directly to the Maréchal de Saxe." (d'Argenson to the spy Aulent, at the camp before Tournai, 13 May).

"Ensure that you are informed of the movements of the troops and the provisions that the enemy may send on the side of maritime Flanders and of what lands at Ostend." (d'Argenson to Angles, lieutenant of the King at Calais, 14 May).

"It is necessary at present to attempt to know what are the troops that they have brought into these places, since their defeat." (d'Argenson to Sieur Thirion, Captain of the Gates at Valenciennes, 15 May).

Henry Pichat

es, lieutenants of the King, ministers at Liége and on the Rhine, as well as spies, sent him numerous reports. These documents give very precise information on the resupply of these fortresses, the preparations for their defense, and the administrative measures taken by the Austrian Government; but it appears more confused according to the indications given on the movements of the Allied troops. Marches and counter marches of the latter appeared most complicated and Thirion, one of the most attentive agents wrote, with most appropriateness, as one will see, "that it is a veritable shuttle that they make."[15]

Invested with supreme command of the Austrian and Dutch troops, Marshal von Königsegg found himself placed in a delicate position by this mark of high confidence. On the other hand, Waldeck "Commander-in-Chief of the Auxiliary Troops of Their High Powers the States General of the Lowlands"[16] possessed formal instructions from the Republic to respect the terms of the military alliance.[17] These terms served for the Prince to refuse to satisfy the requests of the Marshal each time they did not seem reconcilable with the requirements of protocol. Otherwise, the Anglo-Hanoverians reproached the Dutch for having allowed victory to escape them at Fontenoy.[18] During the retreat, some violent protestations were raised in the British ranks.[19]

"Continue your observations and above all pay attention to that which may come from Luxembourg." (d'Argenson to de Robert, lieutenant of the King at Maubeuge, 18 May). As one can see, each of these people received a special mission determined according to his situation and his means of action.

[15]Thirion to d'Argenson, Philippeville, 17 June (Corresp. A.F.).

[16]Waldeck.

[17]Königsegg to the States, Lessines, 6 June (Corresp. G.H.). – The Marshal claims that he made "requests" to Waldeck. This request was frequently renewed."

[18]"It is debated if the Duke of Cumberland had directed these reproaches to General Waldeck for not having acted with his troops when the English and Hanoverians were at grips with the troops of the King." (Thirion to d'Argenson, Philippeville, 21 May). – With a discretion full of implications, the English General Ligonier ends thus his account of Fontenoy addressed to Lord Chesterfield: "Some examples are needed of what I have learned, but it is not our business." [Relation of the Action at Fontenoy (A.L.H.)]. – Lieutenant General Aylva wrote to the states: "I will say nothing of our Dutch troops, they were not engaged except in the village of Fontenoy, but they well could have behaved better." (Relation of the Action of Fontenoy. loc. cit.). Waldeck called a council of war which pronounced chastisements against six Dutch officers who were broken, driven away, or imprisoned [Sententie in de raak den Auditeur militaire Ravens der auxiliaire troupes van den Staat in de Oostenryksche nederlanden eisher in cas criminal op en te gen Appius, Sickenge, etc. (Correspo. G.H.)[

[19]"The English and Hanoverians, upon returning from their defeat, had pillaged the villages and suburbs around Ath. The locals were obliged to close the gates of this city fearing disorders. They have said that butchery had occurred." [Thirion to d'Ar-

- 13 -

Maurice De Saxe's 1745 Campaign in Belgium

"This overexcitement had already caused brawls in the Ostend garrison."[20] Königsegg echoed this unfortunate malevolence: "the difference and the inequality of the troops was too marked that day," he wrote, to La Haye, in the middle of numerous protestations of devotion and respect. After having given praise to the English infantry, he added that some of the soldiers of the Republic sustained with difficulty the fire in the field and that it would be imprudent to expose them to it again. He decided, as a consequence, to employ the immediate reinforcements that the Provinces alone could furnish, by virtue of their reserve, in fortresses.

Königsegg spoke of this to Waldeck. In all his rejoinders the Prince showed that his instructions demanded "equality in the garrisons." The Marshal then called on the good will of the Duke of Cumberland, who was "Extremely equitable in everything." His Royal Highness, most conciliatory, pointed out nonetheless that his soldiers had already reinforced Ghent and Audenarde.[21] He promised, however, "only his troops in case of subsequent need of reinforcements." [22] Could the Marshal reduce the effectives of the solid English infantry without producing a real prejudice to the common cause? He turned the difficulty by drawing from Luxembourg four Austrian battalions, paid for by England, in order to place them in Namur.[23] The army then received a reinforcement drawn from the garrison of that fortress. This maneuver was repeated several times, occasionally a perpetual back and forth motion, a veritable shuttle.

Königsegg foresaw that this "dispute over equality, which assisted the cause of the enemy, would result in the loss of the country and the army." [24]

---

genson, Philippeville, 21 May (Corresp., A.F.)]

[20]"Everyone is, in Ostend, in great consternation; the English complain that they were not well seconded by the Dutch. Several English soldiers have already fought with different Dutch of the regiment that is in garrison there." [d'Aunay to d'Argenson, Dunkirk, 20 May (Corresp. A.F.)].

[21] Königsegg to the States, Lessines, 6 June (Correspo. G.H.).

[22] Königsegg to the States, Lessines, 6 June (Correspo. G.H.).

[23]Königsegg to the States, Lessines, 6 June (Correspo. G.H.).

[24]"I beg Your Excellency to consider our situation with his high luminaries and flatter myself that he will find himself that it is necessary to prefer the common good and press into a dispute over the quality of the garrison. I begged him to make Their High Powers try this maxim, and that the Prince von Waldeck will receive your order to conform to the provisions that are found most appropriate for the betterment of the common causes, in the assurance that I will take all care possible necessary for the conservation of the troops and for the interests of the Republic. (Brief van Königsegg

Henry Pichat

For the moment, this was an inopportune discussion. It presented moreover the serious defect, and to some extent apparently justified that mistrust that already existed among the troops. The defeat had produced quarrels that were not going to be long in degenerating into true hostilities, thus causing new difficulties with the command, which was already a prey to other concerns.

No matter what it was, little by little, this army was reinforced. The English cavalry was remounted.[25] The artillery drew from Brussels the field material destined to replace that which had been abandoned on the Fontenoy battlefield.[26] Waldeck raised free companies some of which had already furnished good service.[27] "The Allied army grows prodigiously." wrote the spy Aulent to the Ministry, on 8 June.

The resistance of Tournai, besieged by French troops, permitted all this reorganization. On 23 May, the city capitulated; but the governor requested orders from La Haye before surrendering the citadel and his garrison. The Powers left the fate of the garrison to Königsegg, who, without delay, sent his instructions to the governor.[28] Considering this fortress as one of the premier fortresses of Europe, the Marshal clearly explained: "to defend the citadel to the last extremity, to not be sparing of either the enemy or of the garrison; during this time, we will gain the time to reinforce our army."[29] Despite everything the citadel capitulated on 12 June. However, as one will see, this prolongation of the citadel's resistance for a month was sufficient. Unfortunately, for the Allies Tournai fell at the precise moment when circumstances were particularly grave. On 6 June, in effect, the first rumor of a possible reinforcement for the Royal Army arrived at the

---

aan den Hr. Raad pensionaires, Lessines, 24 July 1745 (Corresp. G.H.)] The Marshal had already presented a similar request on 6 June.

[25]Thirion and the Angles to d'Argenson, Philippeville and Calais, 10-17 June (Corresp. A.F.).

[26] Thirion to d'Argenson, Philippeville, 12 June (*Ibid.*).

[27]Waldeck, 30 May. – Thirion informed the Ministry, on 10 June, that Waldeck was disposed to raise some free troops: "The General has submitted to the acceptance of the States the names of some officers chosen to exercise command; among whom figured Ferret, a veteran officer of French troops." -- This information is exact, because this partisan gave to the Allies many services for which Waldeck gave him homage. Ferret confirms, in particular, the arrival of reinforcements from Germany in the French lines (Waldeck, 25 June).

[28]Their High Powers to Königsegg, 26 May, Lessines (Corresp. G.H.).

[29] Königsegg to Lieutenant General Dort, Governor of Tournai, Lessines, 29 May (Corresp. G.H.).

Lessines camp. [30] Fifteen days later, the Allied generals no longer doubted the arrival at Tournai and at Maubeuge of a considerable number of French troops coming from Germany. In that way, all the sacrifices made by the Allies since 11 June proved useless. The Pragmatic Army was to maintain its numerical inferiority. In the haste to reform itself, the Allies had imprudently stripped part of the garrisons from their defensive front. This was an aggravating circumstance, at the moment when the fall of Tournai permitted the reinforced French to move forward, without the Allies being able to foresee in which direction they might move. The danger of the situation was quickly revealed. On 25 June, a council of war gathered with Königsegg "to deliberate on what it should do in the existing circumstances." [31] The result of this council was that they resolved to put themselves in defense on the right bank of the Dender at the first alert; but a message from Waldeck to the States[32] and the movements of the troops beginning the next day showed that the weakening of the garrisons of Mons, Namur, and Charleroi raised great concerns. The Allies soon set about returning the troops that they had withdrawn from these fortresses. [33]

At the end of June, there were: At Mons, 6 battalions (4 Dutch and 2 Austrians); at Namur, 10 battalions (4 Austrian, 6 Dutch, some Dutch cavalry, 100 cannon, 80 mortars), totaling 7,000 men; at Charleroi, 1,200 men (48 guns, 12 mortars); at Audenarde, 3 battalions (1 Dutch, 1 Austrian, 1 English); at Ghent, 600-700 English infantry; and at Ostend, finally, about 2,000 men.[34]

As for the field army, it contained, at this time, 58 battalions, and 93 squadrons.[35] It had increased by 12 battalions (9 Dutch, 2 English, and 1 Hanoverian). The comparison of the effective strength with those that Waldeck had furnished in the situation dated 11 May permits us to determine that there were 46,000 effective combatants, maximum, in the Pragmatic Army in line on 1 July 1745. In sum, the Allies had barely repaired their material losses suffered at Fontenoy.

---

[30] Waldeck, 6 June.

[31] Waldeck, 24-25 June.

[32] Waldeck to Grefü er Fagel, of the States General, 25 June (Corresp. G.H.).

[33] Waldeck, 25,26, and 27 June.

[34]Garrisons of the cities besieged by the French in the Austrian Lowlands in 1745 (A.L.H.).

[35] Orders of battle of the Allied army in the Lowlands formed after the arrival of reinforcements which were attached after the battle at the end of June 1745. (Waldeck).

Henry Pichat

Before Fontenoy, the Allies discounted the victory in advance.[36] Defeated after a murderous 8-hour battle, followed by a difficult retreat, they suffered a cruel deception. In leaving the battlefield the Allied army carried the germs of the quarrels whose growth appeared to favor the separation of the very disparate elements that formed it. Provisionally united for the needs of a common cause little susceptible to unleashing the enthusiasm of the Dutch and English, who were already discontented with each other, forming, in reality, the cause of an army fighting to defend the Austrian patrimony and the Imperial ambitions of Maria Theresa, Queen of Hungary, they lacked nothing about which to complain.

Finally, they had little confidence in their generals, for whom, otherwise, the obligation to strictly observe a Byzantine protocol added to the numerous difficulties created by the loss of the battle. The Duke of Cumberland had, for the eyes of the English "all the qualities that one attributed to the great Condé, save for the ferocity." [37] After the battle, however, the Austrians and Dutch only saw in him a 22-year old with little military experience. [38] The judgments of Königsegg (he was 72 years old) and Waldeck, were not more favorable, nor more just, even though they were more or less unanimous.

---

[36]"We had the facility to gain, today, the Tournai plain, to array the army in order of battle and to make such good dispositions that one had every reason to expect a good success all the more since the troops had shown much desire to fight, they advanced with ardor, and they began with some success." (Königsegg to Françoise de Lorraine, Ath, 11 May). – Cf. d'Arneth, Vol. III, p. 413.

[37] "It remains to me to say a word on His Royal Highness, who, by a marked providence, finds himself in good health. The Marshal von Königsegg was surprised by the genius that he found in the Prince for the trade of war and developed a particular affection. [He had a] solid judgment, a rare calmness on occasions where the loss of life is one of the least evils that one can comprehend, an intrepidity without equal, the most brilliant valor, and great activity is the quality which I noted with such surprise in the young Prince during this day. Finally, he had all the qualities that one attributes to the great Condé, except for the ferocity. It is a pity that he will not go far, since Heaven is not obliged to make miracles every day." [Ligonier to Chesterfield (Relations of the Action of Fontenoy, *loc. cit*)].

[38] "These officers said haughtily that they cannot be well led since the Duke of Cumberland, aged 22, had only seen the battle of the Main (Battle of Dettingen), that von Königsegg is better for politics and is always employed there, and that von Waldeck, aged 40 years, has never seen anything." [Robert to d'Argenson, Maubeuge, 16 May (Corresp. A.F.)]. These words were reported to Robert by one of the agents employed in the Allied army's transport. This man had marched with the Allies before and after Fontenoy.

In this way, the existing circumstances imposed on this army a purely defensive role. It could not repair, except with difficulty, the material and moral weakness that struck it to almost impotence. Though the Allies had a larger army, the French, whose success had emboldened them and who were commanded by a man of war whose talents were hardened by numerous campaigns, came to be consecrated by virtue of the brilliant victory of Fontenoy.

Situation of the French Army 1 July 1745.

Placed under distinct commands, the French forces in Flanders and Germany prepared to pursue separate operations. Their respective theaters of operations found themselves too close together and too close to France's borders for these armies to be totally independent of each other. It is, therefore, necessary to know, at least summarily, the events that passed in Germany up to the end of June in order to understand the exact situation of the Army of Flanders at this time.

On 30 March 1745, the French forces in Germany remained spread along the Main, in Swabia, and in the Breisgau. On 1 April, the Court placed them under the sole command of His Serene Highness the Prince de Conti. This general received both diplomatic and military powers, as well as very detailed, but confused, instructions. They ended, however, with precise instructions: "The objective of His Majesty is to be, finally, in a state to prevent and to suspend the election, until the laws of the Empire and the justice of the rights of the Allies of His Majesty can prevail over the intrigues and the seductions of his enemies."[39]

This resolution determined the expectant attitude of this army. Conti could not, in effect, assume the offensive without violating, to his profit, a neutrality that he had as his mission to oblige the Circles and the Electors to respect.[40] Besides, according to the words of the Marquis d'Argenson, Minister of Foreign Affairs, the voting Princes of the Diet "should not become slaves of Vienna except in their misery

---

[39]Instruction for the Prince de Conti (Corresp. A.A.).

[40]"We have assembled, on the Lower Rhine, a superior army to that which the Allies of the Queen of Hungary have united on the Lahn to prevent them from penetrating into the Circle of Swabia and Franconia and imposing on the Ecclesiastic Electors the necessity of subjugating their suffrages in the future election of an Emperor to the interests of the Court of Vienna." (Instructions for the Prince de Conti, *loc. cit.*).

and sustaining their rights supported according to whether they were strong and powerful."[41] The [French] Court hoped that the proximity of Conti would communicate to them sufficient energy for them to conserve all independence; but it also thought that the fear of seeing their territories invaded would assure the sincerity of their intentions.[42]

In any case, the future operations of the Prince depended above all on the fortunes of France's Ally, the Elector of Bavaria, who was always indecisive, and also on the actions of the Duke d'Aremberg, the General of the Queen of Hungary, who occupied the Lower Lahn and a line from Wetzlar to the Rhine, with 30 battalions and 37 squadrons.[43]

At Versailles, one was unaware of all the intensions of this general because he had, in effect, remained immobile in his position since 1 March.[44] Certain strategic considerations and also the hope of obliging Aremberg to unmask his projects, determined the concentration, on the Lahn, of 54 battalions and 68 squadrons forming a new army, called the Army of the Lower Rhine, commanded by the Prince de Conti.[45]

In awaiting the arrival of His Highness, Maréchal de Maillebois completed this assembly, on 15 April. Aremberg left the French in indecision; because, upon the approach of the French troops, he contented himself with establishing himself on the Rhine between Coblentz and Neuwied.[46] The Court thought, then, that the general

---

[41] Rathery, *Souvenirs et Mémoirs de d'Argenson,* Vol. IV, p. 385.

[42]"The object of H.M. is to impose above all on the three Ecclesiastic Electors by making them fear the effects of the ill-will of His Majesty, if they give themselves over to the views of his enemies." (Instructions of the King to the Prince de Conti. *loc. cit.*).

[43]Order of battle: Der unter dem Commando des Herzogs von Aremberg stehenden allierten Armée, April, 1745 [Browne'schees Manuscript (K.A.)].

[44] Aufmarsch der Pragmatischen Armee an der Lahn (*Oesterreichischer Erbfolg Krieg,* VI, p. 502.)

[45] "It appears that the Lower Lahn is the canton most appropriate for the assembly of the army; the resources that one can draw from Hunsruck and the facilities that the Rhine give for transport assure the supply of the force and deny to the enemy the means of establishing themselves in this region and they will be, according to appearances, obliged to withdraw into the Duchy of Berg. There is every reason to believe that the Electors of Trier and of Cologne, seeing in this circumstance their states exposed to total ruin, would be obliged to listen to the voices of conciliation." (Instructions of the King to the Prince de Conti, *loc. cit.*)

[46]"There are, in the conduct of the Duke d'Aremberg, since he approached the Lahn, a sequence of things that are not understood." (Renaud, Minister of the King at Trier, to d'Argenson, Coblentz, 15 April (Corresp. A.A.)].

of Maria Theresa would return to Flanders or attempt a diversion on Alsace, and thought, in consequence, to organize detachment from the troops of Conti's army, in order to oppose these efforts.[47] For the moment, there was nothing there but a project, but it was not long in being realized.

If the attitude of Aremberg remained enigmatic, that of the Elector of Bohemia, however, became bad. Left with the hesitations of Maximilian, Maria Theresa thrust Bathyanyi forward. In no time, the general was master of Bavaria. De Ségur, commanding the small French corps put into the service of France's ally, was obliged to retire, in order to avoid being eradicated, at Pfaffenhofen, on 14 April, by the 25,000 soldiers of the Queen. After a difficult 10-day retreat, these troops stopped on the Necker, where, to collect them, the French Court had hastily recalled those that had previously occupied Austria. [48] On 25 April, Ségur, at the head of 32 battalions and 66 squadrons, was in a position of security, but Maximilian, treating with Austria, abandoned their alliance with France.

The collapse of Bavaria profoundly modified the situation. The French forces now found themselves divided into two groups. However, Conti had also, before him, two adversaries, Aremberg and Bathyanyi, and was unaware if they would act in concert or separately. French impotence did nothing to prevent the release of the hostility of these two electors, who had previously voluntarily observed their neutrality. Besides, this unfortunately justified, and at French expense, the opinions of d'Argenson.

These events did not give Frederick II any occasion to calm his ill-humor, when a new defection was to be feared. The abandonment of Prussia could have disastrous consequences. If this power concluded a peace, Charles de Lorraine would thrust his forces into Germany, before Conti, and leave Aremberg at the liberty to execute the projects that the Court had given him. On 1 May, d'Argenson sent to Conti new instructions inspired by the fears already established and for which the new situation gave a greater force: "Your Serene High-

---

[47]It appears that von Löwendahl had suggested this latter hypothesis in a letter written to the Minister of War. This document bears an annotation, in d'Argenson's hand; "I have communicated this idea to the King; it merits attention, and, to this end, one cannot be too attentive to the movements of the enemy and to be instructed of their designs. It is this that I shall recommend to the commander on the frontier." [Löwendahl to d'Argenson, Sarrelouis, 15 April (Corresp. A.A.)].

[48]These troops occupied Friburg, Brisach, Constance, and the Brisgau, in the name of the Elector of Bavaria, France's ally.

ness shall evacuate from the present location and successively, send a corps, of up to 20 battalions and 40 squadrons, to Thionville and Longwy. It must be in a state to join the army of His Majesty before the arrival of the Duke d'Aremberg, if he marches on Flanders, or to be equally available to pass to the Army of Conti, if d'Aremberg remains on the Rhine and if the Circles declare themselves [against France], or if Prince Charles returns to this river."[49]

One watched, more attentively than ever, the least movements of the duke[50] and the Army of Germany prepared to organize the detachment that the Minister organized; although started their march these troops remained subordinated to the movement of Allies.[51]

In the meantime, the thunderous blow of Fontenoy struck, whose echo immediately rang into Germany. On 14 May, d'Argenson officially informed Conti of the Royal Army's victory, stating that Louis XV desired to draw all the advantages "that should be possible" from this happy success of the 11th. The Minster sent, by the same courier, definitive instructions.[52]

These instructions directed the general to retire on the Main and to move de Ségur closer to him, in moving him to the Lower Necker. The French troops united then could better hold their positions.

[49]New instructions from the King for the Prince de Conti (Corresp. A.A.).

[50] Conti to d'Argenson, Langen-Schwalbach, 7 May (Corresp. A.A.); Esmale, Minister of the King at Liege, to d'Argenson, 19 May (Corresp. A.F.). – The most active agent of this surveillance was Renaud, the French minister to the Elector of Trier. "It turns against me, the bad humor of the Duke d'Aremberg," he wrote to d'Argenson (Renaud to d'Argenson, Coblentz, 10 May (Corresp. A.A.)]. It is necessary to read, in Renaud's correspondence, the savory details that he gives to explain how he counted on escaping the officer and 12 Austrians charged with capturing him. He finished by seeking refuge with the Jesuits of Coblentz. He found there a secure asylum for the Minister of the Most Christian King. The study of this very voluminous correspondence of Renaud leaves one to think that de Broglie had shown, perhaps, too much severity in regard to this diplomat (*Marie-Thérèsa Impératrice*, Vol. II, p. 171). Renaud did not consent to be the victim of the duplicity of the Bishop of Trier and gave proof of a certain generosity of sentiments with regards to this Elector; because he wrote to d'Argenson: "The Austrians are masters of Coblentz, which, at the present moment, the Elector is almost obliged to submit to the law of the Duke d'Aremberg." (Renaud to d'Argenson, Coblentz, 10 May (Corresp. A.A.)]. – "I am very content with this Renaud," said Conti to d'Argenson. It appears that in these difficult circumstances where he finds himself the Prince is an excellent judge." [Conti to d'Argenson, Langen-Schwalbach, 20 May (Corresp. A.A.)].

[51] "The movement of the detachment must not occur until the march of the enemy beyond the Rhine is decided by a day's march forward." [Conti to d'Argenson, Lange-Schwalbach, 8 May-14 May (Corresp. A.A.)].

[52] D'Argenson to Conti, camp under Tournai, 14 May (Corresp. A.F.).

They were obliged to evacuate them, so they moved up the Rhine, "prepared to cross the Rhine only in the last extremity." There was no longer a question, as one saw it, of opposing the Imperial election, nor of delaying it, and the program established on 1 April appeared considerably simplified. Conti was not fooled, when, announcing the coming gathering of the Diet at Frankfurt, on 1 June, he wrote: "this will be a coarse grab of authority, which will greatly alienate the Empire, which will break up. I beg you to inform me if the the King's intention is that one break up the assembly, if he wants that it be declared for him or that one insinuated, in a word what form this step should be given, or if he wants that one to let things progress without interfering."[53] The Court replied that the Prince did not need an army as large as the one he had to play the role imposed on him. France did not have, in sum, on this theater, any one more than Maria Theresa as an enemy, and it sufficed that Conti possessed the forces necessary to not allow himself to be intimidated by d'Armeberg and Bathyanyi. Otherwise, a movement expected impatiently was proper to save appearances and provide a justification for the Royal intentions. D'Aremberg broke his camp, descended the Rhine, and established himself below Bonn.[54] Despite the fact that Conti regarded this retrograde movement as a certain indication of a future junction of the two lieutenants of the Queen[55], the Court put on the show of considering the maneuver of the Austrians as inspired by a care to respect the independence of the Diet. Louis XV profited from this to display the same scruples. Alone, henceforth, the presence of Maria Theresa in Germany opposed the complete evacuation of German soil by the French army.[56]

The Minister pressed the arrival of the detachment requested on the Prince de Conti. He resolved, in addition, to move this corps on the Meuse, "which would be very important," he added "relative to the views of His Majesty; the enemies are seeking to draw back all that

---

[53] Conti to d'Arensen, Lange-Schwalbach, 16 May (Corresp. A.A.).

[54] Baumès, Minister to Bonn, to d'Argenson; Conti to d'Argenson; Langen-Schwalbach, Renaud to d'Argenson, Coblentz, 23 May (Corresp. A.A.).

[55] Conti to d'Argenson, Heidelberg, 26 May (*ibid.*)

[56]"It is in this spirit that the King instructed his ministers in the courts of Germany of his intentions and of his projects to continue the war during the course of this campaign, without bothering on one side the liberty of the German Corps and without suffering, at the same time, that the enemy can boast of the assistance and the facilities that the Princes and the States of the Empire, who support them, can give them to move to attack us. [d'Argenson to Conti, camp under Tournai, 14 May (Corresp. A.A.)].

they can from the garrisons of Charleroi and Mons, to strengthen their army. On the march of the French troops in Alsace, they shall truly be obliged to withdraw these reinforcements so as to not leave these posts exposed." [57] This resolution dedicated the Army of Germany to a definitive inaction and all efforts in the campaign were now directed to Flanders.

Conti immediately put his troops in movement. On 25 May, 4 battalions and 9 squadrons concentrated at Kirchberg; on leaving the following day, they marched on Thionville, under the command of Maréchal de camp Seedorf.[58] That same day 5 battalions and 8 squadrons assembled at Kreutznach, under the orders of Maréchal de camp Relingue, which marched in the same direction. Finally, on the insistent news from the Minister, on 31 May, 11 battalions and 22 squadrons, commanded by Maréchaux de camp Chépy and de Fiennes, left the camp at Flersheim to also march on Thionville. All these officers were to find, in this city, orders giving them their new destination, because part of their strength remained in Mézières, while the rest was going to join the army at Tournai.[59]

On 1 June, the Court designated Lieutenant General Clermont-Gallerande to command the troops assembled under Mézières. Equipped with special powers[60], this officer arrived in the city, on 8 June, and prepared to receive Seedorf there on the 9th, Relinque on the 13th, and the others on the 19th and 20th.[61] But new orders directed him to send to Tournai the 9 battalions forming the infantry of the first two convoys, once they had arrived at Mézières. He then, with all the combined cavalry coming from Germany, was to camp below Maubeuge, until further orders were received.[62] On 29 June, Clermont-Gallerande had 3 battalions and 34 squadrons. He held himself ready to enter the field.[63] For the moment, this general particularly threatened Mons and

[57] D'Argenson to Conti, camp under Tournai, 18 May (*ibid.*)
[58] Conti to d'Argenson, Langen-Schwalbach, 22 May (*ibid.*) This detachment also contained two free companies. It arrived at Thionville, on 2 June. [Seedorf to d'Argenson, Thionville, 3 June (Corresp. A.F.)].
[59] D'Argenson to Creil, Governor of Évêchés, camp under Tournai, 8 June (*ibid.*)
[60]Order of the King to Sieur Marquis de Clermont-Gallerande (de Vault, p. 287).
[61]Clermont-Gallerande to de Saxe, Mézières, 8 June (Corresp. A.F.).
[62] D'Argenson to Clermont-Gallerand, camp under Tournai, 10 June (*ibid.*). – The 3rd Division was taken at the passage at Rocroi by Lieutenant General Boufflers, and moved under Tournai, where it arrived on 28-30 June.
[63] "I think I can be, on the 29th, in a state to make some movement against the enemy fortresses, if you desire." [Clermont-Gallerande to de Saxe, Maubeuge, 22 June (Corresp. A.F.)].

Charleroi, because his detachment had the appearances of an investment corps, which the Royal Army was going to launch against one of
these two places to place it under siege. One had seen, in the previous
paragraph, that this concentration did not escape the attention of the
Allies and that it produced the expected results.

The presence of Clermont-Gallerande's forces gave complete
security to the French frontier. The Court judged it useless, for the
moment at least, to re-established the "lines" constituted by actual
National Guards formed of peasants and organized by Maréchal de
Belle-Isle in 1744.[64]

Clermont-Gallerande established himself, with his 1,800 infantry and 5,000 cavalry, before Maubeuge, under the pretext of defending the frontier and maneuvering on the Sambre. In reality, "he
was to contribute to the operations that the King had ordered in his
plan and would soon undertake."[65] In this manner, the prolongation of
the resistance of Tournai, until 20 June, had favored the reinforcement
of the Army of Flanders.

The Court and Maréchal de Saxe put this delay to profit to
perfect the organization and the state of the army. In order to inhibit
the all too frequent desertion at this time, above all among the free
troops and hussars, it was decided that no one would leave the army
without a passport. This was also a means of watching for the spies
that the enemy, according to the French example, might bring into the
army. The strictest discipline reigned. Contrary to the regulation prescriptions, a number of officers did not camp with their troops.[66] The
King made examples and threatened to remove them.[67] Orders and the

---

[64] "The King did not judge it necessary, at the present, to fatigue the country by a new
re-establishment of the line of the previous year." [D'Argenson to Creil, camp under
Tournai, 23 June (Corresp. A.F.)].

[65] De Vault, p. 48.

[66] This prescription figured in the Royal Ordinance of 20 July 1741 concerning the
discipline of troops in the camps: "Art. 14. All colonels, mestres de camp, and other
officers of the corps shall be obliged to lodge under their tents, even when they find
houses on the ground where they are camping, despite the prohibition against those of
them who found themselves lodged in houses, which might even be occupied by the
brigadiers only in the case where they are on the ground occupied by their brigades.
(Collection of Royal Ordinances.)

[67] "This is not only because the Count de Beauffremont is personally lodged that the
King has determined to make an example, but because he had suffered that the officers
of his regiment are lodged in the same manner. It is for this reason that His Majesty
had wished that the punishment begin with the colonels, who were responsible for
the errors of the officers under them." [D'Argenson to Beauffremont, camp under

Henry Pichat

most severe proscriptions were addressed to the troops coming from Germany. [68] In addition, on the 11, 17, 22, and 28 June, the army "foraged" on the countryside. These operations were carried out pursuant to the "Regulation for all the Campaign, Concerning Order and Police in Forages", established with an eye towards scrupulously avoiding pillage and arbitrary requisitions. Finally, 7,700 prisoners were assigned, on 10 May, to perfecting the works of the fortresses on the Lower Escaut [Scheldt], to repair the roads linking Tournai to Douai, to Valenciennes, and to Cambrai, to fill in the trenches dug during the siege of the citadel of Tournai and finally, to dismantle this work. [69] The city was going to become the head of the line of communications for the supplies of the future army in the field.

In sum, on 1 July, the mobilized French forces consisted of 124 battalions and 213 squadrons divided into two groups: the Royal Army and the troops detached in garrisons.[70]

The first contained: 1° 108 battalions, 167 squadrons, and 100 cannon placed under the direct orders of Maréchal de Saxe.[71] These troops formed a total of 63,608 infantry and 24,000 cavalry present[72], assembled at the camp under Tournai, and ready to march out; 2° the 3 battalions and 34 squadrons under Clermont-Gallerande. Camped

---

Tournai, 23 June (Corresp. A.F.)]. – D'Argenson to Montmorency, colonel of the gendarmerie, camp under Tournai, 24 June (*ibid.*)

[68] "Upon leaving from their camp, these troops went to camp in the vicinity not designated by the Burgermeister or the lieutenants of the King…. They shall march in the best order and with the most exact discipline. A public notice shall be given, upon leaving, at the head of the troops, by which it shall be forbidden to go marauding on pain of death. There shall be three musters per day, one upon leaving, one upon the halt, and upon arriving to control the troops." (Corresp. A.A.). – As the bad weather obliged cantoning the troops instead of encamping them, upon leaving Rocroy, Boufflers made his excuses to the Maréchal de Saxe, Avesnes, 24 June (Corresp. A.F.)].

[69].State of prisoners assembled by the army (de Vault, p. 284).

[70] Brézé, 3 July. – Order of battle of the Royal Army of Flanders and General Situation of the Army of Flanders (Corresp. A.F.).

[71] Ibid.

[72] State of the Regiments of Infantry of the King's Army on 1 July 1745 (*ibid.*) – This situation, destined for the Minister, gives a figure of 63,608 present under arms instead of 72,179 infantry. The difference is the sick and absent. To establish the effective strength of the cavalry, a squadron was considered as containing 140 men and it was supposed that the cavalry had shrunk at the same rate as the infantry. The medium evaluation of the squadron is certainly not exaggerated, because 30 squadrons of dragoons in this army had 150 men per squadron, and some squadrons of cavalry, recently formed, it is true, contained 150 horsemen, while the 22 squadrons of the King's Household, each had 165 men per squadron. (General Situation of the Army of Flanders, *loc. cit.).*

under Maubeuge, these 6,800 combatants were to coordinate their movements with those of Maréchal de Saxe; 3° finally, 10 battalions of militia, totaling 5,195 combatants, were cantoned at Tournai under the command of de Brézé. These troops formed a depot where the field army could draw reinforcements and reserves, as well as, later, garrisons for captured fortresses.

The second was formed with: 1° 5 battalions distributed between Valenciennes, Saint-Omer, Saint-Venant, Aire, Béthune, and Dunkirk, and destined to put the fortresses in Maritime Flanders in a state of defense against a coup de main that might be attempted by an English corps suddenly coming out of Ostend or Nieuport, or by troops landing on the shore; 2° 12 squadrons spread in Landrecies, Givet, Marienbourg, Philippeville, and Tournai to serve along the frontier.[73]

In sum, an army of around 90,000 combatants, enthused by a great victory, strongly organized and well commanded, was ready to put itself on the march.[74] It found itself in an excellent material and morale situation to undertake the offensive which the command had decided as its objective and for which it had prepared the means.

The Campaign Plan

At the moment when the Royal Army was ready to more forward, barely 30 kilometers separated it from the Allies. The lightly wooded terrain did not present any great obstructions and extended between the two adversaries. Cut by numerous insignificant water courses, it did not offer, in this season particularly, any significant obstacles that made the Pragmatic troops, weak and demoralized, safe against a decisive attack. It appears, therefore, that the French had nothing to do but to march against the Allied army to attack and beat it.

This solution, whose circumstances were imposed on a general of the 19th century, could not be adopted by Maréchal de Saxe in 1745. The tactics of the time, and the strategy that resulted, did not permit imposing a battle on an enemy to prevent him from executing a definitive retreat.

---

[73] Brézé, 3 July.
[74] The free companies are not included in this figure. There were five such companies, however, furnishing a total of 150 horsemen and 500 infantry.

Henry Pichat

However, in the actual state of things, it was impossible to embrace all the Flanders theater of operations by the separate corps, combining their respective movements, in light of bringing the Allies to battle, because of the numerous fortresses that were spread across the terrain.

It is necessary to consider, finally, that the armies of the 18th century did not have a road network sufficiently developed to facilitate their marches. In Flanders, there were very few good, paved roads. Beyond them, the traveler only found mediocre and narrow roads. When winter came, these latter were impracticable, except during a freeze, and during the good season, as well, the frequent rain in this region made them very bad, and the movement of artillery or convoys rapidly destroyed them.[75] Finally, the columns of troops were very long, and the roads were frequently not direct. The generals preferred, sometimes, to march the soldiers cross-country, be it to shorten the march, or be it to avoid using the bad roads. They confined, then, to their chief of staff (maréchal général de logis de l'armée) the job of reconnoitering the chosen itinerary and executing, as necessary, the repairs necessary to facilitate their passage.

The march formations further augmented these difficulties. The troops pulled along with them their baggage.[76] The "menus",

[75] The weakness of the roads obliged the King to promulgate, on 1 December 1746, an ordinance to prohibit the use of two wheeled carts. "His Majesty had recognized in the last campaigns that the wagons, forges, and other two wheeled vehicles, supporting only on two points the loads with which they are charged, destroy the highways and the roads that they travel, to the point where, in a single march, they are rendered impractical and that the draft horses are placed one behind the other such that they so lengthen the files of the train that the escorts cannot adequately cover them over all their length, and wishing to prevent the disadvantages that arise from this, all vehicles with two horses shall be eliminated from the army. (Collection of Royal Ordinances).
[76] The Royal Ordinance of July 1716 fixed the number of vehicles allocated to the troops: three wagons for a battalion; one for a squadron, while they are enroute, for the transport of their baggage and the sick.

A new regulation of 8 and 15 April 1718 augmented the numbers by one wagon per battalion and one for every two squadrons.

Many complaints were raised, so a new ordinance was issued on 5 December 1730, which changed the numbers: five carts or wagons drawn by four horses for a battalion; three with four horses, per regiment of two squadrons; four with four horses for each regiment with three squadrons (Collection of Royal Ordinances.)

These vehicles were paid before the departure of the troops, at the rate of 20 sols per hours, except in Flanders, where one paid 30 sols. It was prohibited to load them with wine, foodstuffs, or merchandise, except for what was appropriate for the troop. The commissioners of war and the Intendants alone who could modify the figures: "in the case where the troops were charged with new clothing or large distri-

wagons bearing their lame, ammunition, food, as well as materials indispensable to troops that never cantoned, followed the brigades. The "heavy" were relegated to the tail of the column, containing personal vehicles, as well as the multiple and varied baggage of the officers and the troops. A turbulent throng of the army's valets[77], women, children, merchants, and animals accompanied it as well. The provost and his escorts could barely maintain order. One figures easily that the presence of Louis XV with the Army of Flanders was not of a nature to reduce these impedimenta.

These endless convoys loaded down the columns and frequently their assembly gave rise to serious disorders and considerable

---

butions [of materials] that could not be distributed before the departure."

The Royal Ordinance of 20 July 1741, summarized those of 1 April 1703, 25 February, 1 April 1705, 15 February 1734, 8 April 1735, and gave the following authorizations: "the Marshals of France, the lieutenants general, the commander-in-chief or commander of a corps or troops are authorized the number of heavy vehicles that they judge appropriate; but in regard to the other officers, each lieutenant general, not commanding, shall have no more than 30 horses or mules including those allocated for the hauling of three wheeled vehicles, but they may not have a larger number of horses or wagons. Each maréchal de camp, 20 horses, including those drawing two wagons. Each brigadier, colonel, or mestre de camp, 16 horses, including a single wagon. Lieutenant colonels, captains, and other officers may not have any personal wagon, or no more horses than those for which they may draw forage, unless the King grants it. His Majesty permits the officers of each infantry battalion and each cavalry or dragoon regiment, to have common wagons, in the train of the battalion or the wagon, drawn by four good horses.

"Each cavalry or dragoon regiment and each infantry battalion may have, in addition, a sulter or a wagon on the condition that it is drawn by four horses. If there are other sulters, they may only have saddle horses. It is, finally, permitted that each regiment have a baker with a four-horse wagon. In order to further lighten the train, on 29 June "chairs are forbidden to all officers, including colonels." (Godefroy).

These figures permit one to calculate the train of an army containing about 30 lieutenant generals, 50 maréchaux de camp, 70 brigadiers, 108 battalions, and 167 squadrons. One must not forget to add the baggage of Maréchal de Noailles, those of the Minister, those of Maurice de Saxe, and the 25 officers of his staff, and finally those of the King and his domestic household.

[77] The Royal Ordinance of 27 December 1743 prohibited "all officers of troops to use soldiers as valets." It recalled the officers to the strict observation of the Royal Ordinance of 13 July 1727, to abstain from presenting their valets in the reviews of the commissioners of war, employing effective soldiers of their companies for their personal service, in the capacity of domestics and finding the means to relieve them from the ordinary service which falls, by this means, on the shoulders of their comrades; to which it is necessary to provide, as much to make service equal among all the soldiers of the company as to maintain the spirit of honor that is not very compatible with domestic functions."

Henry Pichat

delays.[78] For all these reasons the march of the army could only be very slow.

Besides, it was easy to foresee that the Allies would endeavor to avoid contact. Also, at the cost of such slow and difficult marches as those of the French, it remained free to displace at its pleasure over a land that was its own. It held Ghent and Audenarde on one side, Ath, Alost, and Dendermonde on the other; as well as Mons, Charleroi, and Namur; to speak only of the important places. Camped on the Dender, ready to hold up French operations on this river or the Scheldt, they could also make French operations in Babrant by the Haine, the Sambre or the Upper Scheldt very difficult.

If the Allies, supported on their fortresses, equivocated for two or three months, perhaps more, the Maréchal de Saxe could hope to follow them with the single hope of seizing a favorable occasion of a false maneuver [by the Allies] to defeat them again. To this end, he risked using his numerical superiority and of finding himself again, at the end of the summer season, with a fatigued army before an enemy who may have been reinforced.

The Court resolved to draw the best part of the excellent situation they held at the end of June. It decided to pursue the execution of the offensive program laid out in December 1744 by the Maréchal de Saxe, and to assure the army of a position that was more favorable each day, by taking from the Allied fortresses that would serve them as

---

[78] R.A. recounts the following upon his departure from the Leuze camp (See Chap. II.) "I made arrangements to leave the camp. I was on horseback at 3:00 a.m. but I could not leave Leuze until around 5:00 a.m. It was still necessary to risk being crushed by the horses of the King's baggage and the prodigious number of valets who, in the support of the people of the domestic household of H.M., thought themselves excused from obeying the order that had been given them not to pass by Leuze. After having pierced this flood, I joined the camp, of which a part was not yet at the rendezvous, because it had been stopped by the crowd in Leuze. While one was entangled there, the columns of crews slipped by and the route, which was to hold the camp, was occupied so that we were obliged to change roads and we arrived around noon at the camp where were to have been at 6:00 a.m." – "The marches of the Royal Army, at the beginning of July, were marked by many disorders. Maréchal de Saxe was obliged to remedy them. The march order was always regulated with great precision. The army was broken into four, six, or eight columns. Each of them received a reconnoitered and very detailed route, which used the road network with the greatest degree. The provost force was augmented; one demanded of them the strictest service. Despite all the cares the army could barely march between 12 and 15 kilometers per day, above all at the beginning of the campaign, where the time of march varied between 8 and 16 hours."

much as cover as points of support. By this means, the French troops increased their field of operations. Besides the events accomplished from the opening of the campaign assured the means of developing with more fullness the initial plan of campaign they had developed, to augment their number of conquests, and to grow, at the same time, the activity of French diplomacy to the point where negotiations might begin for a general peace.

The Pragmatic Army remained free to oppose the French movements. If it risked battle, it would be destroyed by a sure blow, because of its weakness. As a consequence, it became absolutely negligible, on the condition that, maintaining their forces, the French retained the great numerical superiority that they had acquired on 30 June. In sum, it was pointless to attack the Allies. It sufficed, so to say, to menace it.

What should be the objective of this offensive? Two solutions presented themselves, following that the Royal Army operated by its right or left, that is to say, to the east or the north of Tournai.

The northern French frontier found itself freed and Maritime Flanders was put in a state of defense against surprise, so one could think of enlarging the breach opened in the barrier by first taking Mons and then Charleroi. It was difficult to act better in French interests.

The report of December 1744[79] had already exposed the disadvantages of such a manner of acting: "Mons taken will give us the ability to live in the Brabant, but nothing more, and still, we would not have to go much beyond this conquest, because being obliged to leave part of our forces on the coast, the enemy would render our subsistence more difficult. I want to defeat them, which will not carry us much further and it will always be necessary to bring back our troops to our country to spend the winter or to still undertake bad sieges, which will only require us to establish garrisons in them. As for the enterprise against Charleroi, I will not speak about it. The advantage of this place, if it is followed by fortunate successes, would give us nothing but a few more facilities to undertake other sieges." The events had not subtracted anything from the force of these considerations. Extension towards the east, would produce the loss of our numerical superiority, by the obligation of garrisoning Tournai, Mons, Charleroi, and of sending, into Maritime Flanders, the troops necessary to reinforce

---

[79] Commandant Colin has authenticated and published this important document in his work *Les Campagnes du maré*chal *de Saxe*, Vol. III, *Fontenoy*, Supporting Documents, P. 20.

the manpower [there] as what was left there was barely sufficient to guarantee this province from a surprise attack. To the contrary, an offensive from the north offered only advantages. The report of December 1744 very fortunately brought them out: "The conquest of Tournai facilitated for us that of Audenarde and Ghent, which would assure us all of the left bank of the Escaut [Scheldt] to the sea, and Ostend would become useless to the English." Under these conditions, the French could remain grouped and cover Maritime Flanders.

The sights of the Marshal were wider still. "If we hold this position," he wrote, "the enemy would only be able to establish their [winter] quarters behind the Dender and we will make them very jealous during the winter and multiply so greatly their expenses, that they will be short of many things; because it is necessary for them to think about Ostend, Antwerp, Brussels, and all the little fortresses along the Dender; however, they do not dare to strip Mons. In putting 15 or 20 battalions into Ghent during the winter, to hold the head of our [the French] quarters, we can put 40 squadrons on the wall that closes the canal that runs from Bruges to Ghent and the Escaut [Scheldt]. We have all the land of Waës before us which we can force to contribute [to us] during the winter, as far as Antwerp and Brussels, and all the region situated between the Dender and the Escaut. Our communications from Lille and Tournai to Ghent is easy and covered by the Escaut [Scheldt]. It will pass along good roads and from the Lys, advantages which no other position in this region can give us. These operations for the campaign of 1745 appear to me to be sure, easy, and commodious for all the means and assure us an important position for the coming winter. All these reflections have decided me to undertake a siege of Tournai in preference to the other places, and consequently, to move the war on the Escaut."[80] There is nothing to add to this picture.

The excellent situation of the army, at the moment of the resumption of operations, cannot but confirm the Marshal in his manner of seeing the situation. Maurice was, therefore, going to move to the Escaut; that is to say, capture Audenarde and Ghent; and afterwards, "one will see."[81] Did he perhaps plan on taking Ostend and the forts along the Dender? These later conquests would remain subordinated to the resistance offered by the Allies. No matter what happened, at the

---

[80] Colin, *Les Campagnes du maréchal de Saxe*, Vol. III, *loc. cit.*

[81] De Saxe to Chevalier Folard, Borst, 18 July (Corresp. A.F.).

end of the campaign, the Marshal could winter his army between the Escaut [Scheldt] and the sea.

Maurice was to still better carry out these hopes by the grace of his vigor, his skill, his prudence, and also, it should be said, to the inertia of the Allies.

CHAPTER II
The War on the Escaut [Scheldt].

Uhlan and Dragoon of the Voluntaires de Saxe by *Philibert-Benoit de La Rue*

The pursuit of the war on the Escaut [Scheldt] demanded, initially, the capture of Audenarde, and then that of Ghent. To this end, according to the tactics of the time, the general had to constitute a detachment destined to alternatively pursue the sieges of these two places, under the cover of an observation corps establishing a screen before the Pragmatic Army. About 20,000 men sufficed to capture Audenarde.[82] If Maréchal de Saxe made such dispositions, the Allies would not risk attacking the 70,000 Frenchmen that remained with the Royal Army, after the formation of a siege corps. The final success of French operations was, therefore, not in doubt.

The Allies, camped at Lessines, remained, nonetheless in a position to cause some difficulties. At the least movement in the direction of Audenarde, it could occupy, before the French, the heights dominating this fortress on the right bank of the Escaut, block the investment, then, finally obliged to retreat, they could reinforce Ghent.

---

[82] See, on the subject of the importance of a siege corps, Colin, *loc. cit.,* Vol II, p. 219.

Maurice De Saxe's 1745 Campaign in Belgium

They covered, thus, the evacuation, on Dendermonde or Antwerp, by the Escaut, and the mass of provisions gathered in that place. In sum, not only an important capture could escape the French, but the anticipated sieges could be completed during the good season.

It appears, however, that he only had to extend his hand to seize Ghent and its 700 defenders, as only 20 kilometers separated it from the extreme right of the Allied army.

Admirably informed by his spies, Maréchal de Saxe was not unaware that, since 19 May, the Allies had established, behind their camp, the bridges necessary to cross the Dender.[83] Absolutely useless to assist Audenarde, this disposition could, if needs be, be regarded as the first of the preparations for a rapid and secure retreat, behind the cover of the river, with an eye towards covering Mons. The precaution taken by the Allies witnesses even more their care to avoid a new battle. "They act according to what you do,"[84] reported a French spy, and the attitude of the confederates betrays their indecision.

The siege of Audernarde could not begin before the Pragmatic Army had moved away from it. The Marshal did not doubt that by making a movement on Ath, that is to say, by threatening Mons, it would not determine the enemy to move closer to this latter city. He sought a means to post French troops between the Allies and Audenarde; after which he would detach his siege corps. This maneuver, not presenting any serious difficulty, would be crowned by a certain success. Maurice only had to march, openly, against the Allies to oblige them to move away or to give a battle that was impossible to sustain. Besides, the cramped nature of the free ground before the French made this maneuver obligatory. There only remained to choose the best the manner to carry it out. Up to what point, in effect, would the Allies beat a retreat? Would they still waver on left bank of Dender a while longer; or, at the first feint, would they immediately seek the shelter of the right bank? In the first case, there only remained the resource to achieve the operations as just stated. If, to the contrary, the Allies adopted the second solution, the Marshal would pursue a plan that was as simple as it was bold. He would not change anything in the covering role reserved to the Royal Army; only the operations that it was to mask: an energetic and rapid blow directed first on Ghent and then the vigorous siege of Audenard would be inverted.

[83]Waldeck. – Corresp. A.F., May-June.
[84] Thirion to d'Argenson, Philippeville, 16 June (Corresp. A.F.).

The paradoxical character of such a project constituted one of the best guarantees of its success. If, in effect, the suspicions of the Allies were not aroused, could they envisage an immediate attack on Ghent, which it hoped well to see fall as late as possible, while admitting that even the general situation did not improve? Besides, such a certainty, among the Allies, alone could explain why a garrison of 700 men defended this fortress.

As a result, the Royal Army acted to threaten Mons, by moving its 90,000 men on Leuze, by the first march. That same day, with an eye towards simulating the preparation for the investment of Mons, Clermont-Gallerande camped at Binch. If the Allies responded to this feint by a retreat sufficiently pronounced so as to abandon to the Royal Army a field of operations sufficiently wide, it would maneuver in order to isolate Ghent and Audenarde. Once this result was obtained, the expedition, secretly prepared with care, and destined to capture the two places, would leave at the moment when the Allies could not oppose de Saxe's projects. Thus, suddenly and in a short time, the French could cut the direct communications between England and the Continent and take from the Allies two fortresses, with considerable provisions. One cannot doubt that such an operation, after the victory of 11 May, would produce considerable repercussions as well as important results.

On 23 June, Waldeck reported to the States General the surrender of Tournai, adding: "that the moment appeared to have come to prevent the French from doing another thing." [85] This appeared easy, on the condition, however, that the fortresses were in a state of defense.[86] The council of war, convened two days later by Königsegg, shared this perception, because Mons, Charleroi, and Namur were immediately re-manned. It is certain also, one would soon have the proof, that to assure the security of Audenarde, the Allies preferred to move their right wing closer to this position, rather than reinforcing

---

[85] Waldeck to the Greffier Fagel, Lessins, 23 June (Corresp. G.H.). – He added: "If our troops cannot appear again in the field, the French having their elbows free would do much harm before things change face and, thus, one gains the time necessary for a powerful diversion in the Empire, that one can reasonably expect, which puts us in a position to undertake something in our turn."

[86] "I am persuaded, that after a costly siege as that of Tournai, the French will not tie themselves up with another place of consequence, at least that they do not find it stripped so that it cannot defend itself and that we are not, at the same time, in a state to blockade it with our army." (Waldeck, to Their High Powers, Lessines, 23 June (Corresp. G.H.)].

the garrison and they thus avoided reducing their forces in the field. Waldeck, otherwise, affirmed the urgency of these dispositions: "The enemy," he said, "will make his attack after he has taken the arrangements necessary for Tournai, or he will do nothing." Were Cumberland and Königsegg convinced by this assurance, where the General of the Republic mixed his singular naiveties with his unfortunate illusions? Only the result of the events permits one to suppose this. It appeared soon, at Lessines, that he had resolved to not lose an instant in moving closer to Audenarde. In addition, the reports of the free companies returning from expeditions had exaggerated the importance of Clermont-Gallerande's detachment and confirmed the concentration of this corps under Maubeuge.[87] The reconnaissance also reported also that the French were cutting down all the woods situated between this latter place and Mons; from the advance posts of the Pragmatic Army, one saw French pioneers repairing the roads to the North and South of the highway running from Tournai to Leuze.[88] Everything, in a word, announced the pending march of Maréchal de Saxe, when, on 28 June, he executed a forage under the same conditions as those of 11, 17, 20, and 24 June. Was he then going to remain in his camp for four more days? The Allies resolved to put to profit this delay that seemed certain. On 29 June, they reconnoitered a new camp. They occupied it the following day, their spies persisting in affirming, despite appearances, the imminence of the French marching out.[89]

Remaining on the left bank of the Dender, the Pragmatic Army marched by its right, crossing the Ogy Stream at the Accrène bridge and over the neighboring pontoons that it had established upstream, then camped in two lines from Grammont to Lessines, facing to the west. A corps, called the reserve, and formed almost uniquely of English troops, occupied this latter position.

This closing on the Escaut did not appear to favor the secret intentions of Maurice de Saxe. It could not, however, inspire any fear in the Marshal, because the threat that the French were making on Mons prevented the Allies from accentuating their movement. How-

---

[87] Waldeck, 25 June.
[88] Waldeck, 26-27 June.
[89] "The army marches in three columns, by the right, as it had camped. The 1st and 2nd lines formed into two columns, the artillery, and the baggage in the third." (Waldeck, 28 June). – The detachments from Leuze and Herchies occupied Enghien, that of the de Ligne Château, Ollegnies,and that of Hamaide remained there as well as those of Scoorisse and Everbecq. A detachment of hussars occupied Ninove and an Austrian company occupied a post at Ogy. (Waldeck, 30 June).

ever, such a withdrawal showed that the suspicions of the Allies could be easily awakened on the same side where it was important that they disappear.

No doubt, the French march had been prepared for a long time.[90] Was the departure immediately decided on by the recent movement of the Allies or simply by the arrival, on 30 June of the last reinforcements coming from Germany?[91] Be that as it may, the movement order for 1 July suddenly raised the indecision that reigned among the French troops. The forage of the 28th, in effect, had announced to everyone "four days at least of residence[92]", while the others, more skeptical, ensured that it would be the last and that, perhaps, the army would not consume it.[93]

On 30 June, the troops made some preparations. The heavy baggage of the army, as well as a great part of that of the King, parked on the glacis of Tournai, at the Lille Gate[94], that is to say, in the proximity of the cobblestone road to Leuze, which they were to follow the next day. That evening, one assembled at the park all the army's sick and disabled.[95] Some brigades camped a distance away on the left bank of them Escaut came, to close in, and bivouac on the right bank of the river.[96] Moreover, all these movements were barely possible for a few days only, the baggage of the Allied garrison captured at Tournai did not leave behind its columns until 24 June.[97]

On 1 July, the Royal troops shook themselves out. The advance guard, stripped of all its baggage, departed at 3:00 a.m., formed with 6 regiments of dragoons (30 squadrons) commanded by Maréchal de camp Chevreuse and some camp selection parties and new guards, under the orders of the *maréchal de camp de jour*. The army formed in five columns. During the duration of the siege of Tournai, the corps had camped on the line of circumvallation, according to needs. The

[90]Godefroy, June.
[91]"On 20 June, the 1st Division of troops from the Rhine arrived; on 22 June, the 2nd Division arrived; on 23 June, all the artillery came to the park or the place d'arms at Tournai from the Lille Gate to prepare to march forward." (Godefroy, 4 June).
[92] Pons, 28 June.
[93]Bréze, 29 June.
[94] Bréze, 29 June.
[95]Bréze, 29 June.
[96] "On 30 June, the Uhlans (Saxe-Volontaire) passed the Escaut, the Gendarmes, who were in Bachy, the Gardes at Cisoin, Bourghelles, the Grenadiers à cheval were at Pont-à-Tressin, Baisiez, Esplechin, moved to camp at Calonne." (Godefroy).
[97] "24 June, Thursday. The baggage of the garrison began to march out at 6:00 a.m. They were gone by 8:00 a.m." (Godefroy).

brigades were, as a result, disorganized. In addition, the arrival of the troops from Germany necessitated the restructuring of the order of battle. The command resolved to profit from this circumstance to re-establish everything. As a result, and by exceptional measures, each of the five columns formed at a unique assembly point and, to move there, all the corps received a special movement order fixing their departure time and the itinerary to follow.[98]

The signal was given by the *lieutenant général du jour*[99] and the columns departed. The wings of the march (columns 1 and 5)[100] contained 48 and 40 squadrons respectively; columns 2 and 4 were formed exclusively of infantry (40 and 45 battalions respectively). The center contained 31 squadrons and 15 battalions followed by what we shall call, from now on, the convoy; that is to say, the treasury, the King's baggage, those of His domestic household, of the Princes, of the headquarters, of the general officers not employed, the troops of the column, and the dragoons; the artillery and the hospital closed the march. As the third column alone contained the heavy baggage, it was provided with a rearguard.

The army utilized the road network almost exclusively and did not move away from it except to find better or shorter roads, across the fields. The cobblestone road from Leuze to Tournai, the axis of march, was reserved for the center column.

The French worked to avoid the disorders that the assembly of this formation, quite complex, would surely produce. On the 30th, all of the majors reconnoitered the itineraries imposed on their respective corps.[101] The Marshal reiterated the formal order to depart exactly at the prescribed times and to march the soldiers in dense ranks. [102] Each brigade was to be followed by its baggage, leaving an officer to follow the wagons immediately behind the troops. [103] Finally, the regiments were to avoid cutting each other off and massing as much as possible.[104]

---

[98] Brézé, 3 July.

[99] The services of the day consisted of: a lieutenant général and a Maréchal de camp, as general officers. It is necessary to add a brigadier, a colonel, a lieutenant colonel of infantry, the commander of infantry pickets, and the same service for the cavalry.

[100] Columns were always numbered, conforming to the order of movement, in the natural sequence of numbers, from the right to the left in the direction of march.

[101] The order of march for 1 July. (Brézé).

[102] Brézé, 1 July.

[103] Brézé, 1 July.

[104] Brézé, 1 July.

Despite everything, the movement was long and difficult, above all because of the complicated formation of the five columns. "A movement made the night before had made the march easier and more pleasant," said a contemporary judiciously.[105] It would have been preferable, in effect, to reform the brigades and to re-establish the battle order before departing. Among all the movement orders that the staff of the Marshal established during this campaign, none offered as confused and complicated a document than that of 30 June. Though the King standing on a height near Leuze watched the different columns moving out, extending, and camping on the plain, the last elements of the army did not rejoin it until very late in the evening.[106]

During the demonstration for the offensive on Mons executed that day, which was even more complicated, Clermont-Gallerande moved from Maubeuge to Binch where he camped.[107] Leaving his camp at Haumont, he crossed the Sambre at Boussois then crossed Erquelines and Merbes-Sainte-Marie. He had his march scouted by a detachment of cavalry which drove back into Mons a reconnaissance that had come from that fortress and spread alarm. At the same time, the Marshal sent about 300 of his light cavalry (Beausobre Hussars and Saxe-Volontaire Uhlans[108]) through the Brabant.[109]

They Royal Army camped, according to an instituted formation, to the south of Leuze and on the highway from Ath to Tournai[110],

---

[105]R.A'.

[106] R.A'. – R.A. states 11:00 p.m.; Brézé only says "very late."

[107] "Our arrival greatly shook up the city of Mons, where the cartwrights and the blacksmiths were busy constructing the gun carriages." [Clermont-Gallerande to d'Argenson, Binch, 2 July (Corresp. A.F.)].

[108]The Saxe-Volontaire were a cavalry troops recruited by Marshal Maurice de Saxe based on his ideas for mobile warfare. It consisted of dragoons and uhlans (lancers). The troops were organized and dressed according to the taste and ideas of the field marshal which were armed with carbines and lances.

[109] M. de Saxe to Folard, Borst, 18 July (Corresp. A.F.).

[110] In general, the army camped according to a more or less unchanging formation in two lines, the cavalry on the wings, the infantry in the center of each of them, and the troops were deployed according to the order of battle. Thus, are explained the denominations of infantry of the right-hand side of the 1st and 2nd line, left wings of cavalry of 1st or of the 2nd lines, etc., employed in the movement orders. The result of this formation is that one can constitute the columns simply by breaking the lines in the chosen direction without changing, each time, the enumeration of the brigades composing each of them. Moreover, the use of the assembly points is eliminated. The movement order gains in conciseness as well as clarity and the installation in the camp becomes more rapid. The French seldom varied from this habit, except in the case of absolute need, or when, considering a lack of time, the maréchal général des logis had

facing to the north in three lines. The camp was formed with infantry on the left and cavalry on the right, the second of the two wings of the cavalry and a center of infantry, the third, finally, forming a sort of reserve, was formed only of cavalry. The artillery and the baggage parked to the north of the highway, facing Leuze. The dragoons and the Guards Brigade, in front of the village, covered the various quarters where people were lodged in Leuze. No tactical necessity had imposed this exemption from the usual rules. A contemporary wrote: "one knew that he had a camp to occupy at Leuze; one was content in this knowledge, and one found it so rich in men that the day following the march, the maréchaux des logis asked again: 'where will we put ourselves?'"[111]

The Allies could not be unaware of the French movement. As of this day however, the Allies had been fooled regarding the true the object of this march. Having received information that "French marched on the Allied army[112]," Waldeck sent reconnaissances along his front[113] and Cumberland reinforced the posts at Everbecq and on the road from Audenarde, on 1 July. Indecisive, and fearing of being surprised, the Allies moved part of their heavy baggage on Ninove, on 2 July, at 2:00 a.m., and then put themselves under arms.[114] At noon, they were still there, when their spies informed them that the French were disposing to move their camp from Rebaix to Wannebecq. The Allies did not doubt that Maréchal de Saxe went to seek combat rather than besiege of Mons. As a result, Königsegg prescribed an immediate retreat to the right bank of the Dender: the feint attempted by Maréchal de Saxe had completely succeeded.[115]

Beginning at 2:00 p.m., the Allied movement ended at nightfall. After having marched around Grammont, the Allies supported their right on the Dendra, against that fortress, on the heights which, to the east, dominated it. Their left was established near Gaimerage. The swampy course of the Macq, or the Enghien Stream, covered the front of the Pragmatic Army and the Dender their right flank. It is appropriate to say that the evacuation of the left bank was not absolutely complete, because even if the posts at Hamaide, Everbecq, Schoorisse,

not sufficiently reconnoitered his emplacements.
[111]R.A'.
[112] Waldeck, 1 July.
[113] "I executed a reconnaissance on Meaux." (*Ibid.*).
[114] Waldeck, 2 July.
[115] Waldeck, 3 July.

and Lessines were withdrawn, the reserve corps still camped before Grammont, on the road that linked that fortress with Lessines.[116]

A few advance posts still held out in Accrène and the bridges upstream. The fear of a battle had inspired this rapid retreat to one of the strongest positions in all of Flanders.

Ath found itself uncovered. Königsegg judged it indispensable to replace the 1,200 men that he had recently drawn from the fortress' garrison. During the retreat and two other times, he had begged Waldeck to throw a few troops into Ath.[117] This was a rational request, since the Dutch effected their movement into the region neighboring the city. With an astonishing inappropriateness, the Prince refused to assent to the Marshal's request, invoking the requirements of the military convention. "I will," he said, "supply the Marshal, by the same messenger, to have the goodness to consider that the Dutch already have a battalion there and that the English do not have a soul there." Königsegg resigned himself and, on 3 July, only, some English troops were sent to Ath.

Besides, provisionally at least, Wurmbrandt, the Austrian governor of this city, could deploy 3,000 soldiers. He could not assemble artillery on the ramparts, because, two months later, it was not there yet.

The information on the faith with which the Allies had so precipitously withdrawn was exact, but incomplete. On the 2nd, in effect, the French army had not moved. The order to march had, it appears, already been given.[118] "The Maréchal général de logis made his representations; they were heard and he was given a day to prepare to lead us to Rebaix."[119] In the meantime, French reconnaissance detachments[120] brought in news during the evening of the Allied retreat. It was there that a very important event occurred to suspend the

---

[116] "The right wing, in two columns, marched by its right and crossed the right at Ockerzeele and Nieder Boulaer. The left wing, also in two columns, crossed the river by the bridge at Accrène and over a bridge upriver. The artillery and baggage, in part, crossed at Grammont. The post occupying Lessines served as a rearguard for everything. The march of the infantry and the cavalry was correct and good, but the artillery became entangled with the English baggage in Grammont and could not reach its camp until the next day." (Waldeck, 2 July).

[117] *Ibid.*

[118] "The order had in some fashion been given." (R.A'.)

[119] *Ibid.*

[120] "In the evening, one saw appear some parties of French on the Lessines heights; I withdrew the Hamaide detachment." (Waldeck, 2 July).

march of the army. Besides, in effect, the Allies seemed to abandon to the French all the ground between the Dender and the Escaut. It was indispensable that the French assured possession of it. From the morning of 3 July, 12 companies of grenadiers sent to the de Ligne Château found it empty and occupied it.[121] There was soon no doubt that the Allies did not have "so to say, not a cat on the left bank of the Dender."[122] The Marshal could, therefore, from that moment, complete the last preparations for the projected expedition on Ghent, which was then still not suspected.

Far from raising the indecision of the Allies, the march on Leuze had inspired in them the new fear of the imminent shock of the 90,000 French soldiers. To prevent that the Allies, return to the left bank of the Dender, it sufficed to continue the threats that had determined them to place themselves in a state of defense on the right bank of the river. Otherwise, out of fear of moving too far from the Royal Army, the movement on Mons could not be pushed too far.

The Marshal was then going to maneuver to mask Audenarde and Ghent by giving his movements the appearance of a march to battle. In awaiting to see, if the movement to prepare the second part of his operations had arrived, that of the execution remained some time off and the Marshal was to continue to conceal his projects, in order that it would be impossible for the Allies to interfere when they finally realized what was happening. Maurice wrote up the last orders given to Lieutenant General Count Löwendal chosen to capture Ghent and to besiege Audenarde. Dated from Leuze, on 3 July, his instructions remained confidential.[123] All the army knew of the Allied retreat behind the Dender; everyone understood the necessity of assuring it; but no one suspected the preparations that had been completed on the 3rd.

One can say that the operations against Ghent began on 4 July, the day where the French troops resumed their march. At 3:00 a.m., the advance guard formed encampments and new guards, commanded by the maréchal de camp de jour. He was given 50 workers, with their tools, because the columns were to move along a long road across the fields and cross a number of streams.[124]

---

[121]Brézé, 3 July.

[122] R.A.

[123] The movement order for the 4th, found in the *Journal* of Prince de Pons, convened with Maréchal de camp de Chevreuse, attached to Löwendal, during the evening of the 3rd. The instructions received by de Chevreuse remain unknown.

[124] Movement order for the 4th (Pons).

Henry Pichat

Upon the sounding of a long roll of drums, Maréchal de camp de Beauffremont led away two regiments of dragoons.[125] This detachment moved from Irchonwels to Bouvignies, outside of range of the cannon at Ath and to mask this place so as to cover the right flank of the army's march. The Grassins, excellent light troops, following, moved on Lessines from where the last posts of the Allies had retired after a sharp skirmish.[126]

The brigades marched in six columns. Because of the lack of a paved road in the zone of march, the heavy column was divided into two on 1 July and separated from the convoy of troops that had preceded it that day. The general formation remained appreciably the same. The right and left wings contained 48 and 40 squadrons. The center contained two columns of infantry (2nd and 5th) with 44 and 41 battalions flanking the 3rd (31 squadrons and 15 battalions) as well as the fourth (heavy column). All the trains containing the heavy baggage received rear guards.

The cannon at Ath, on the right, as well as the large woods at Hamaide and Oedenghien on the left, limited the zone of march to Villers Saint-Amand, Montbreau, Bouvignies, and Rebaix, one on side, as well as to the road from Hamaide Mainvault on the other. The terrain available presented a width of about 4 kilometers. The obligation of passing the entire army through this choke point as well as the intention of utilizing the bridges and the existing crossing points over the numerous streams explains why the army, on this day, marched more or less completely cross country.

Not surprisingly the disorder occasioned by the departure of the innumerable teams clogging Leuze[127] was more rapid than 1 July. The troops began camping at 2:00 p.m.[128] The quarters of the King were established at Rebaix; that of de Saxe at Papignies, covered by three brigades of cavalry and the Grassins remained at Lessines. A contemporary recounted that "this time again we were in the same difficulty as at Leuze, because of the disproportion of beds to men seeking a place to sleep." No documents permit us to appreciate the

---

[125] Septimanie and Beauffremont Dragoons, in total 10 squadrons, or 1,500 men. These troops did not rejoin the main body until 7 July. They camped, on the 5th, at Rebaix, resting and resupplying on the 6th, left on the 7th and allied with the main army that same day. (Brézé, 7 July).

[126] Waldeck, 4 July. – The evening of this day, the advance posts of the Allies were finally to abandon Accrène and Boulaër.

[127] See Chapter I, footnote 1, p. 14.

[128] R.A.

value of this criticism.[129]

This march confirmed the Allies' information and fortified his suspicions. The Allies seeing the hope grow that they had conceived of seeing the French commit the imprudence of attacking them in their strong position, did not move. On the 5th, the Marshal accentuated the threat in moving to the Ogy stream (also called the Aurence Stream), with his right supported on Lessines.

On that day, a strong advance guard formed of 20 companies of grenadiers, the new guards[130] (1,000 fusiliers) and 1,000 horse, commanded by Maréchal de camp d'Armentières, assembled at 3:00 a.m., to the north of Papignies. It left to take up position at Ober-Boulaër, masking Grammont and covering the right flank of the march of the French troops. It was also to execute the first demonstration on the Allies' right flank. A reserve corps of 31 elite squadrons (Maison du Roi, Gendarmerie, Carabiniers), led by Maurice de Saxe, in person, left Papignies behind the advance guard and established itself on the Ogy Stream, guarding the crossing points and held itself ready to assist Armentières if the Allies came out of Grammont in force. The convoy preceding the army was to park at Wannebecq. Lightened of their heavy baggage, the Royal troops, moving in five columns, marched out in their turn. The right wing consisted of 48 squadrons and the left wing consisted of 20 squadrons, followed by 20 battalions. The intermediary columns (2nd, 3rd, & 4th) were formed solely of infantry and contained 29, 32, and 27 battalions respectively. Camping parties marched at the head of each column. They stopped when they reached the low heights on the right bank of the stream and awaited the order to set up camp.[131]

An aide-maréchal des logis of the staff of the Marshal had guided each column during its march. The itineraries made the greatest use of the road network, whose resources were further increased by the Allies during all the time that they had occupied these regions. One did not deviate in any way to use the bridges at Gomenpont and Drimpont; the time necessary to construct others was lacking. As for

---

[129] No sketch of the camp was found in the War Archives. Pon's Journal, all the same, gives the plan of almost all the campaign camps.

[130] In the category of "new guards" it is necessary to include those of the King, of the army, and the general officers. These troops were always placed under the orders of a *maréchal de camp de jour* and marched, in general, as a support for the camps. The old guards, under the orders of the *maréchal de camp de jour*, were used to form rearguards or forming special posts destined to cover a march or guard bridges.

[131] March order for 4 July (Brézé).

baggage, they only overloaded the departure of the army. Formed in the same number of columns as the troops, they followed the same itineraries. The King's baggage and that of different quarters assembled at Rebaix where they waited for the order to move out and all the convoys received rearguards, charged with assuring their security and with maintaining order. Maréchal de Saxe gave the order to camp as soon as Armentières reported that the Allies had not come out of their lines and that they had recalled their reserve corps, remaining below Grammont since the retreat of the 2nd.

The army formed in two lines markedly parallel running from Lessines to Flobecq, and covering its front with the Ogy Stream, following the usual formation. The artillery was in the center and the cavalry was on the wings of each line. A cavalry reserve prolonged the second, to the south of Lessines, as far as the Dender. The Grassins remained in this place.

The provision distribution center was effectively peaceful.[132] The criticism, not long before so severe, said only that "this time, everything was well conducted with a few little incidents close to the number which occur when it becomes a crowd."[133]

The Allies were content to fortify positions previously established at Enghien and to also occupy Viane, the crossing point over the Marcq. They projected, finally, to modify the basis of their camp and they prepared, the following day, to execute a light movement, when their spies came to them and informed them of the French approach. On the 6th, the Royal Army had resumed its march at daybreak. It had successfully cut the Allies' communication with Audenarde and Ghent, because on the 7th, the necessities of resupply and also fatigue had immobilized the troops.

To execute the projected movement, the French troops first had to cross the Ogy Stream, wheel to the right and finally, stretch out their left, sufficiently turning the Allied flank supported on Grammont. This was a very delicate maneuver, because no detail could escape the Allies who dominated all the plain from the heights of this little fortress. They could, with a sure blow, become very dangerous if, disabused suddenly and recognizing the danger, the Allies fell on the flank of the French march. Maurice worked to take from his adversaries all inopportune suspicions. Already the demonstration made the day before by Armentières had give the Allies the illusion that

---

[132] Brézé, 7 July.
[133] R.A'.

the French were about to attack their very strong position at Grammont. The immobility of the Allies demonstrates their certainty that they would "receive us there."[134] To further fortify their hopes, on the 5th, in the French camp, the plan to march to battle the next day was announced.[135] The persistence of the Allies' spies had informed them, from the 1st, that the French were going to give battle, no doubt finding their source in the constant renewal of this misleading boastfulness. The following day, finally, the Marshal took such positions that a battle appeared imminent.

At daybreak, d'Armentières left with 40 companies of grenadiers, 40 pickets (2,000 muskets), 1,000 horse of the Maison du Roi, and the Grassins.[136] With these elite troops, he resumed his positions of the previous day and the Grassins opened a fusillade, appearing to begin an attack on Grammont. At the same time, the entire army began its real maneuver.

According to the preparations of the Marshal, his troops formed into three columns. The first line of the camp became the right wing of the march, the second the center, and the reserve, camped under Lessins, became the third line, the left wing, with the baggage train.[137] Each column crossed the Ogy Stream at a special point, the first at the Accrène bridge, the others over two pontoons situated upstream from that bridge. However, these three crossing points found themselves very close to one another and the first line of the camp masked everything. To free their access, the army executed a movement well chosen to confirm, at the same time, the hopes of the attentive Allies. The troops of the first line formed in column by the right, then the tail doubled to the right of the head so that the two pontoons were cleared. To cross the stream, these troops then had only to execute an about face and break by their left which stood at the level of the Accrène bridge. The second line, by a similar movement, doubled into two parts, to the south of the first line established as we have seen, and left free the access to the last pontoon. In its turn, finally, the third line formed in column by a left face, then advanced, passing

---

[134] "They flattered themselves that they would receive us there." [M. de Saxe to Folard, Borst, 18 July (Corresp. A.F.)].

[135] "The King ordered to distribute powder and balls to the troops. The Prince announced that he would march against the enemy the following day." (D'Espagnac, *Histoire du maréchal de Saxe,* Vol. II, Note p. 93).

[136] March order for the 6th (Brézé, 7 July).

[137] By this disposition it was that the column that had the greatest distance to cover and found itself farthest from the Allies, contained nothing but infantry.

the others, until it reached the height of the crossing. Because of the proximity of the three bridges to each other and their proximity to Grammont, the army found itself massed in five columns, at a distance of only 4 kilometers from this little fortress, which is to say, under the eyes of the Allies.[138]

It would have been difficult to better simulate the preparations for an attack. The Allies expected the first French movements. An entrenchment dug around the Allied right was manned by three English battalions and some cannon; two batteries were established at Boulaër. A third took up position at the Grammont position, on the heights[139], and a fire began, which was completely inoffensive. During this time the French columns crossed the stream, and finding themselves, beyond the bridges, the *aides-maréchaux des logis* to guide the march[140] advanced into the plain curiously followed by the spyglasses and "reflection telescopes" of all the Allied staff, which was gathered on the height. The news from the French camp, the attack of the Grassins, the force of Armentières' detachments, and the Marshal's movements were intended to give the Allies the suggestion that the anticipated attack was coming. However, as it was already 6:00 a.m., the Allies still had not understood all these "marches and countermarches."[141] Turning to Schlippenbach, brigadier of the Dutch cavalry, Königsegg asked him his opinion. "I believe," responded an officer being interrogated, "that the Maréchal de Saxe is presenting, thus, his army before our army to make it appear that he plans to attack while he slips a corps of troops past to capture Ghent, Bruges, etc.," – "I believe, on my faith," said Königsegg, "that you have judged it well."[142] No other document confirms the authenticity of this exchange, of which Schlippenbach is the only source. Waldeck reports that the Allies remained under arms until 2:00 p.m., watching the French columns march and camp, in the evening, from Ogy to Everbecq.[143] The Prince added no commentary to this observation. It appears, however, that Schlippenbach's words were of a nature to cause a certain agitation. Otherwise, Cumberland himself, to whom the fate of Ghent could not be a matter of indiffer-

138   Brézé and R.A. give all the details of this maneuver.
139   Waldeck, 6 July.
140   "They led the columns by the roads that they had reconnoitered. In order to be able to repair them, if it was needed, at the head of each column marched 10 artillery artisans with a tool wagon and 150 workers." (Brézé, 7 July).
141   Schlippenbach, 29 June-10 July.
142   Schlippenbach, 29 June-10 July.
143   Waldeck, 6 July.

ence, decided only on the 8th, to reinforce it. For these reasons, the inspiration of Schlippenbach appears suspect; still, it must have been difficult for the Allies to retain their doubts about what was happening.

Be that as it may, after a very slow march[144], the French troops camped towards evening, between Everbecq to Chartreuse or Abbaye-aux-Bois, near Sainte-Maria Liedre, in two lines, facing the Dender and, it appears, in a most irregular manner. Contemporaries do not limit their criticism of this action. According to one of them, a Royal fantasy and the ingrained habits of disorder, already manifested by the baggage, were the principal cause of this long march;[145] a second chronicler gives his criticisms in a different order.[146]

By the movement executed, on the 8th, Ghent was definitely isolated. Assembling at 4:30 a.m., the advance guard marched out at 5:00 a.m., containing the new guards and the camping parties. Behind them, the baggage, like the rest of the army, moved in three columns. After having crossed the Swalme, the baggage parked and waited to enter camp.[147] From the right to the left the columns of the army con-

---

[144]R.A. – R.A'.

[145]"Already, on the 5th, the King, all his suite, and everyone had the pleasure of examining with glasses and telescopes, the enemy camp. This act of kindness done, one resumed the march for Chartreuse. The delay caused by the King resulted in his column encountering another column. His Majesty was obliged to stop for a long time. A second, similar incident occurred shortly later. The baggage, departed as it could, was also interrupted in its march. The result was that the King, on horseback at 5:00 a.m., arrived at Chartreuse at 5:00 p.m. Many people had to do without dinner and the beds of the King and the Dauphin remained in the mud. His Majesty slept on straw in Chartreuse. Everything had united to produce the strangest and most horrible camp that had ever existed." (R.A'). – "7 July sojourn, the King slept on straw having wished that the artillery precede his camp." (Godefroy).

[146] "There were, in this march, two great errors, the first that the bridges placed over the Orgy were too close to one another and too far to the right, which caused the second and the third lines to fall into some confusion, a route too long since after having crossed these bridges these two columns were obliged to move up stream as far as they had marched downstream. The second error was that in the direction of the second column, which instead of moving beyond the first moved between two parts of that line. The third did the same behind the second, such that the marching army found itself doubled into five lines which made it very long and very fatiguing." (R.A.). – It is interesting to present this criticism, because it establishes that the necessity of massing the army below Grammont did not appear to either the anonymous chronicler, nor to the lieutenant generals commanding each column. All the maps of the time mention the existence of a bridge of Ogy. Nothing prevented the Marshal from using it if he had preferred to adopt a disposition perfectly indicated by the circumstances. Otherwise, the army, still not very supple, had maneuvered badly.

[147] Bréze, 9 July.

tained 28 squadrons and 13 battalions; 48 squadrons and 44 battalions; 48 squadrons and 47 battalions. The troops broke by their left in the order where they had camped. It appears that the departure did not occur until about noon.[148] At 11:00 a.m., all the old guards, under the orders of the maréchal de camp stepped out, formed in line before the Paricke mill facing the Allies.[149] They then followed the center column after it had passed. Armentières who had remained at his post under Grammont, since the 6th, formed the general rearguard of the army. The center followed the direct road to Audenarde and the right wing moved along badly traced roads. The Everbecq Woods obliged the left wing to close on the center, and it was obliged, because of the lack of better options, to march seven kilometers cross country. When camping, the army supported its right on Trois-Censes near Seglsem and its left on the bend of the Swalme, near Borst, where the King had his headquarters. The Marshal established himself at Velsique. A narrow stream, with steep banks covered the front of the army like an entrenchment.[150] The Royal troops would rest in this camp until 28 July.

This last march was the occasion for new criticisms. "His Majesty and his army went to Borst with the same inconveniences for which I have so frequently reproached that multitude of men, animals, and baggage," they said.[151] "The march that led us here," wrote others, "is one of the worst one had ever seen. The army arrived there between 9:00 p.m., and 4:00 or 5:00 a.m., the following morning, although we only had 2 leagues to cover. I do not know who was at fault."[152] The author adds to these lines a very detailed and not very

[148] "The 'general' [long drum roll] was sounded at 3 o'clock; the army marched at noon." (Godefroy, 8 July).

[149] No document gives any reason for this formation. One might suspect that it served as a curtain to conceal the march as the army turned its back to the Allies, and it would be hard to cause them to continue to expect the hoped-for battle. The account left by Waldeck seems to support this hypothesis: "The enemy began to march at 4:00 a.m.; but having perceived that some of our battalions were striking their tents, and this was done without any order, they stopped until 11:00 and camped towards the evening." (Waldeck, 8 July). – "One finds in this account all the details of the march: departure of the advance guard at 4:00 a.m., the formation of the old guard in line of battle before Paricke, followed by their departure behind the third column of the army. It is otherwise doubtful that the raising of the tents of a few enemy battalions was an action sufficiently important to stop our troops. The reason for which they left at noon appears to be, no doubt, that before marching out de Saxe wished to be certain that the enemy would not move that day."

[150] Bréze, 9 July.

[151] R.A'.

[152] R.A. The departure of the Chayla detachment, on 9 July, soon furnished confirma-

- 49 -

engaging description of the Borst camp. He ends it with a reflection that one was not surprised to find after such an edifying lecture: "the day before the attack," he wrote, "d'Argenson spoke quite correctly in saying that one had never seen such a bad, ridiculous, or impractical camp as that one."[153] The chronicler saw nothing in this camp but an "emplacement to refresh the army." and, once again, the true reasons, which one had determined the choice of the Marshal, escaped him. Maurice himself furnishes other reasons, though very judiciously however easy to guess at: "I came," he said, "to place myself in this camp, which was very advantageous, because I had fears that while passing Audenarde, the enemy did not move to place himself there, which would have cut my communication with Lille and Tournai and would have forced us to write the second volume in the battle of Audenarde, or at least we would have missed Ghent."[154] Otherwise, other appreciations are different: "the camp is beautiful, well placed on a raised and abundant plain, it appears to me to be very secure," wrote the Prince de Pons.[155]

The true objective of these marches remained uncertain. After the capture of Tournai, the siege of Audenarde appeared certain to be "our principal goal."[156] The march on Leuze created a certain perplexity. Was the Marshal going to attack the small Fortress of Ath? "This meant nothing."[157] Would one besiege Mons? "That was a subject of great risks."[158] On 4 July, however, the investment of Audenarde no longer appeared in doubt;[159] however, on the 9th, one did not see on the side of this fortress any investment nor siege position,[160] indecision redoubled in the French camp. Certain suspicious, however, began to appear. "Did one not have designs on Ghent, stripped of troops and ammunition, and whose population was seditious and numerous?" wrote the Prince de Pons.[161]

---

tion of this information.

[153] *Ibid.*

[154] Maurice de Saxe to Folard, Borst, 18 July (Corresp. A.F.).

[155] Pons, 9 July.

[156] Pons, 2 July.

[157] *Ibid.*

[158] *Ibid.*

[159] "By his retreat behind the Dendre, the enemy abandons to us all that is between that river and the Escaut; we are masters of our operations; we continue to march tomorrow to take from the enemy his communications from Audenarde, von Löwendahl must make the investment from the other side of the Escaut." (Pons, 4 July).

[160] *Ibid,* 9 July.

[161] "I have suspicions that two days will clear this up." *Ibid.*

Henry Pichat

It appears, however, that distrust had still not touched the spirit of the Pragmatic generals, despite the oracle of Schlippenbach, prophet of the Grammont Chapel. In effect, it was only during the night of the 6th, that Cumberland sent his reserve corps on Alost and an English brigade to Nivone.[162] The following day, the Allies sent their heavy baggage to Brussels. The Allies remained in the most complete calm until at midnight its spies went to Königsegg.[163] "All the French grenadiers are marching on the Pragmatic Army," they said. Had the attack vainly anticipated by the Allies the day before finally arrived? They ran to their arms and, when day arose, they saw the French columns marching on Borst.[164] Obviously, they had to abandon their hope for a battle and opened their eyes on a particularly disturbing state of affairs. The urgency to reinforce Ghent struck, but bad news succeeded bad news quickly convinced the Allies that it was too late. Cumberland sent reinforcements to Ghent on the 9th; the 10th, he learned of the bloody check of this effort, as well as Löwendahl's march on Ghent, and on the 11th, finally, he learned of the capture of the city by the French.[165]

Such is, briefly related, the model, simultaneously simple and perfect, of the offensive tactic. Maurice de Saxe had accomplished his mission. Löwendahl would now fulfill his. The troops charged with executing this second part of the plan, the most delicate, did not escape the attention of the general who had prepared it.

Completely arriving at Leuze, on 11 July, the French Army did not leave in a mass. Four regiments of dragoons (3,000 men commanded by Maréchal de camp Chevreuse) left Leuze immediately after the departure of the French troops, on the 4th. Escorting Löwendahl, they crossed through Tournai and encamped that evening at Espierres. The following day, the four regiments of Royal Grenadiers[166] (2,251 combatants under a maréchal de camp and Brigadier d'Hérouville) arrived from Tournai with 400 volunteers led by Méric.[167] At 2:00

---

[162]Waldeck, 6 July.

[163] *Ibid,* 7-8 July.

[164] *Ibid,* 8-9 July.

[165] Waldeck, 9-10-11 July.

[166] It appears that these regiments were a recent creation. "One has formed, this morning, three battalions of militia grenadiers. There are four. The colonels receive 12 livres per day, the lieutenant colonels 10, the aides-majors 3. D'Hérouville commands them." (Godefroy, 2 May).

[167]Captain Méric, of the Piémont Regiment, commanded, during this campaign, a corps formed of volunteers that rendered great services and showed many of the qual-

pm., on the 8[168], the Méric Volunteers[168] found themselves in ambush between Deynse and Ghent, on all the roads and paths that went to that city along the two banks of the Lys. They stopped all traffic towards Ghent until it was too late to bring useful news of Löwendahl's approach. That same morning, the gates of Courtrai and the Harlebecq bridge were also masked so that no news would reach to Audenarde.

Leaving Espierres, in the morning of the 8th, Löwendahl and his dragoons moved to Courtrai after having covered 15 kilometers. Then covering the 30 kilometers more that separated them from Deynse, they stopped in that city until nightfall. They finally departed to cover the last 15 kilometers, in such a manner as to reach Ghent around 11:00 p.m., joining the Méric partisans. Thus, Löwendahl attacked the fortress with less than 4,000 men, of which the most part, mounted it is true, had just marched 60 kilometers. These troops alone undertook the action, because the Royal Grenadiers, leaving after the dragoons, had to stop for two hours at Courtrai and slept at Deynse, after covering 45 kilometers. It was before day on the 9th that they resumed their march to join Löwendahl, who they would find in the city, if his attack had succeeded. These instructions, which did not apply if, "nothing happens that changes the prepared operation," and, out of prudence, one took all precautions for resupplying the troops did not prevent an immediate departure, if one found oneself in the obligation to postpone the operation for a few days.[169] In this case, then, Löwendahl would set for himself the details of the operation "as dictated by his prudence and his control."

From his entry into Ghent, "it would be possible to return to the Imperial Gate, where the Marshal would send a detachment during the morning of the 9th."[170] It seems that one can deduce from these lines that these last troops were to be in front of the place, during the morning of the 9th, and not that they would leave the Borst camp on this date. How otherwise could one explain the injunction to "do what is possible" to open the Imperial Gate? If the detachment left Borst only during the morning of the 9th, one could not arrive before the city until late in the day and if Löwendahl had made himself master of it

ities of light troops.

[168] Instructions for von Löwendahl, Leuze, 3 July (Corresp. A.F.).

[169]"General Löwendahl would take bread for all the troops, on the 7th, to be distributed to the 12th inclusively. De Brézé would provide him with the bread and biscuit, on the 9th, in a convoy to feed the troops from the 12th to the 16th. (Instructions for von Löwendahl, *loc. cit.*).

[170] *Ibid.*

from the morning, this general would have had no difficulty in opening the Imperial Gate. No document informs us of the Marshal's intentions. It is probable that Maurice had assigned to this party the role that it would later play. If, in effect, Löwendahl's attack succeeded, after having been joined by the Royal Grenadiers, this general would have 6,000 men, more than sufficient to hold the new conquest without needing to assign to them this detachment. One can admit that Maurice may have desired to create a diversion favorable to Löwendahl's attack. This is, otherwise, what was produced 48 hours later, by the surprise implemented during the night of the 10th/11th.

In the report addressed to the Marshal after the affair[171], Löwendahl had no illusion about this postponement. He would, however, not have failed to justify it if he was responsible. His *Mémoires* do not explain the reason why. As Löwendahl started from Espirres only on the 10th, he received, therefore, on the 7th, at the latest, orders to postpone the operation, of which no trace has been found. Brézé writes that it must not be seen here anything more "than the result of the disorder so common in war."[172] This explanation is insufficient, coming from the plume of a chronicler who was always precise and well informed, which permits one to think that the causes of this delay were completely by chance.

To research this, it is necessary to consider that, on 7 July, the French army remained in the camp it had entered the previous day.[173] It appears that the fatigues occasioned by the marches of the 1st, 4th, 5th, and 6th, and the need to resupply the troops, determined their immobility. In these conditions, however, the army could not find itself at Borst until the 8th. It is true that, in any case, it would be placed under cover before Löwendahl tried his coup. If it was of little importance to the Marshal's plan to complete the surprise of Ghent that day or if it was essential that the Marshal's maneuver was completed the day before or the same day as the surprise of Ghent, the delay of one day in the arrival at Borst was not without consequence on the marching out of the detachment, because it could not leave before the 9th, at the earliest. In addition, on the 7th, Audenarde was masked and, as

---

[171] Löwendahl to Maurice de Saxe, Ghent, 11 July (Corresp. A.F.).

[172] Brézé, 12 July.

[173] Maurice de Saxe writes in his letter to Folard, already cited: "I have rested, on the 7th, because of [a lack of] bread" and Z ... (Zambault), chevalier of one of the Ordnance Companies of the Gendarmerie, relates in his *Campagne des Pays-Bas par le Roi en 1745*, p. 119: "On Wednesday, the 7th, we rested because of fatigue and to await the baggage which had not yet arrived, as well as that of the King."

of this moment, nothing prevented the preparation of its investment. The command considered this, because, on the 9th, Löwendahl was going to send a convoy to Denyse that the instructions of the 3rd do not mention. One can thus conclude that it is in the delay of the arrival to Borst and in a new organization provoked by this circumstance that one must search for the causes of the 48 hours' postponement imposed on Löwendahl.

Completely disillusioned, on the 8th, as we have seen earlier, the Allies attempted to reinforce Ghent. After having rapidly assembled about 7,000 men in Alost, Cumberland gave command of this force to the Austrian Lieutenant General Mölck, on the 9th, with the mission of reaching Ghent that same day.[174]

Two roads led to this city. One was an excellent highway, about 25 kilometers long, to the right of the Escaut, and the other road, not as good and longer, was on the other bank. The total absence of all communications of the French with the left bank rendered the second route perfectly sure. Mölck, however, chose the first. This resolution, the late departure from Alost, and the total absence of cannon in his detachment, indicates that he expected to complete his mission without obstacles. He only had to succeed.

Barely installed at Borst, Maréchal de Saxe prepared to dispatch the detachment foreseen in his instructions of the 3rd. Profiting from this circumstance, he ordered it to construct a bridge at Quatrecht, in order to establish communications with the left bank of the Escaut and to completely isolate the Fortress of Ghent, whose capture had been resolved upon, even if the coup de main was checked.

Lieutenant General Du Chayla assisted by Maréchaux de camp Graville and Souvré, received the command of this detachment, consisting of two brigades of infantry (eight battalions) and three of cavalry (24 squadrons), forming an effective strength of 4,500 infantry and 3,000 horsemen, maximum, supported by 20 pontoons and 20 Swedish cannon.[175] Be it that du Chayla did not receive his instructions until the morning of the 9th, or be it that one desired to give some respite to the troops fatigued by the march of the previous day, the

[174] Waldeck, 9 July. – Cumberland had sent the reserve corps to Alost as well as an English brigade to Nivone from the 7th. For a reason that remains unknown, these troops remained assembled in this latter city. On the 8th, the Allies decided to reinforce Ghent, reiterated the order for them to go there, and attached three new battalions to them.

175 Translator: Swedish cannon are light guns attached directly to and operating with individual infantry battalions.

departure, initially fixed for 5:00 a.m., did not occur until 10:00 a.m.[176] A very strong force of Grassin arquebusiers, forming an advance guard, marched out early in the morning and took up position on the highway from Alost to Ghent, upriver from the latter city, to cover du Chayla's right flank. Maréchal de Saxe was not unaware of either of the force of the Allied detachment in Alost, nor the Allies' projects. He informed du Chayla, that Mölck had not yet moved when the French troops left, Maurice and his lieutenant correctly supposed that the march of the French detachment would not be disturbed and that Mölck, when he attempted to enter Ghent, would find the French troops installed so as to block the road to him. They had only about 16 kilometers to cover to reach Melle where they were to encamp.

At 10:00 a.m., the companies of grenadiers, camping groups, new guards, pontoons, artillery, and the Normandie, Laval, and Crillon Regiments, as well as all the cavalry followed by the baggage, shook out and marched, according to a contemporary, "as if they are going to march from Paris to Versailles."[177] This disposition appears even worse since the French column moved over a narrow road not far from the Allies. Near 10:00 a.m., a patrol of Grassins suddenly encountered a patrol of Allied hussars at Westrem.[178] It charged them, taking a few prisoners which it sent to Borst and put the rest to flight. Grassin sent a report of this action to du Chayla, adding that he would remain in observation, but that he saw no other Allied troops on the

---

[176] The second hypothesis appears more likely. R.A. recounts, in effect, that the Crillon Infantry Brigade, ready to leave only at 5:30 a.m., requested a rest, which was granted it. The troops had not entered the camp until very late. The infantry of this detachment contained the Crillon Brigade (3 battalions of the Crillon Regiment and one of the Laval Regiment) and that of the Normandie Regiment (4 battalions).

[177] This march order furnished by R.A. Brézé presents it thus: the cavalry, the artillery, the pontoons, and the two brigades of infantry. If, possibly, this last version is in accord with the phases of the combat, whose account follows, this appears possible. One does not comprehend, however, that the cavalry would be in the head when moving over an already bad road, which it would have rendered impractical to the infantry and also that, in these conditions, the column would have taken nine hours to cover about 16 kilometers.

[178] According to R.A', the Grassins were already in Melle and its vicinity the night before. The act is not implausible, since these excellent troops frequently operated a day's march in front of the army. In this particular case, one can barely give credence to R.A'.'s assertion if one recalls that the Grassins had taken part in the demonstrations made against Grammont on 5/6 July. There is no indication that Armentières had any more of these light troops under his command when he formed the general rearguard during the march of the 8th.

road from Alost.[179] At exactly 10:00 a.m., this news could not have been more than two hours old. Unfortunately, as one will see, when Grassin saw a change in this situation, he found it impossible to inform du Chayla.[180] Whatever it may be, the information carried by Grassin's messenger did not trouble du Chayla. Du Chayla continued to march in all security, leaving the lead of his column to be taken by the grenadiers, the camp parties, and the new guards, in order that this detachment could "mark out" the camp with Graville, the general officer of the day.

Situated between the Escaut and the Alost-Ghent highway, about 6 kilometers upriver from the city, the small village of Melle finds itself in the middle of a vast, swampy plain, covered with numerous hedges and copses that blocked the view. Six hundred meters in front of the first houses on the side of Alost was a small priory (a Benedictine abby), surrounded by walls and ditches, forming, with the neighboring castle, two very solid posts that blocked the space between the Escaut and the highway. After having moved along the side of the priory, the highway crossed, by a stone bridge, over the Goutrode Stream. This stream filled the abbey's ditch with water. Above this crossing, the water course pushed into a very thick wood, cutting the terrain to the south of the highway.

Upon arriving at Melle, Graville made his preparations. It appears that they were very unfortunate, given the configuration of the terrain and the information provided upon his departure by Maréchal de Saxe. Graville decided, in effect that, forming the right of the first line, the cavalry should establish itself on the banks of the river, pushing its head up to the branch. The Crillon and Laval Regiments extended this troop, camping under the cover of the priory and the little castle. The Normandie Regiment, in the second line, covered Melle on the side of Ghent and installed itself on the northwest edge of the village, between the highway and the Escaut. Du Chayla, finally, lodged himself with a cavalry guard in the large castle, forming, to the south of the highway, an extension of the Normandie Regiment's camp. The artillery and baggage parked on the same side, halfway between the Goutrode Stream and du Chayla's lodging.

It would have been preferable, it appears, to mass all the French infantry on the side of Alost, the side from which danger was

---

[179] R.A. *loc. cit.*
[180] Brézé, 12 July.

most likely to approach.[181]

Be that as it may, according to these dispositions and in awaiting the arrival of the column, Graville posted in the priory and in the castle the two companies of grenadiers as well as the new guards from the Crillon and Laval Regiments. He sent those of the Normandie Regiment to Melle, to the cemetery and to the exit of the village facing the Escaut. Fifty masters[182], finally, occupied the edges of du Chayla's lodging, watching the ground and the highway on the side of Ghent. They would have been more useful guarding the bridge over the Goutrode that Graville imprudently left with only the grenadiers in the priory to defend. They could also have connected this general officer to the Grassins that were in Westrem; then, the surprise that occurred later would have become impossible.

At noon, Mölck left Alost, 150 hussars forming his advance guard. He then came in person, escorted by 350 Anglo-Hanoverian cavalry and followed by the infantry. The rest of his cavalry and baggage brought up the rear of his column.[183] This lieutenant general brought with him about 7,000 combatants. Put to flight, as we know, by the Grassins, the point of his advance guard had informed him of the presence of French light troops; so that he arrived at Westrem, towards 5:00 p.m., where the Austrian general was pleased to still find them there. He counted, in effect, to benefit from this imprudence of the Grassins to capture them; but he had to fight with a strong party. Falling back quickly, the Grassins threw themselves into Massenem Farm, which was surrounded by walls and a ditch full of water; they barricaded themselves in it and defended themselves vigorously. For more than an hour, the Allies attempted to force their position, but not having any cannon, they were forced to renounce the effort and resume their march.[184] There was already some recklessness in this as it represented a loss of precious time, night was coming, and it was 10 kilometers to Ghent. Mölck then committed another grave error by leaving the Grassins behind him without leaving some troops to contain and watch them. The reputation for boldness, justly merited, of these excellent troops demanded at least these precautions. The

[181] "Du Chayla was advised that he might not have an affair with more than 5, 000-6,000 men. The word of a marshal of France served to guarantee against all events in this regard. It was otherwise assured that he would have nothing to fear from Ghent." (R.A'.).

[182] Translator: The horsemen of the heavy cavalry regiments were called "masters."

[183] Waldeck, 9 July.

[184] R.A. – R.A',

Austrian general took none and he would soon have cause to regret this imprudence.

The French infantry had marched so slowly that the head of its infantry only began to debouch onto the Melle Plain a bit after 7:00 p.m. Covering the 16 kilometers in 9 hours, the detachment was not in a hurry and if the Grassins had not delayed the Allies a long time in front of Massenem, du Chayla would only have arrived in time to seize the Allies' baggage. It would have been impossible for him, in any case, to prevent Mölck from throwing himself into Ghent with the major part of his reinforcements. Otherwise, no messenger arrived to hasten the march of the French troops, since Grassin, then completely invested in Massenem, had not been able to send a single messenger. Already the artillery and pontoons had parked, guarded only by two companies of grenadiers and 150 fusiliers, their ordinary escort, when Mölck's advance guard appeared. The 150 hussars that formed it quickly crossed the undefended bridge. Their arrival produced chaos. The French soldiers had little expected to see the Allies as they engaged in the thousand details of setting up camp. There was a certain commotion that is easy to understand, but it lasted only a short time since the French troops brought themselves to battle with a spirit of initiative, a rapidity, and a remarkable ardor. The grenadiers of the priory immediately lined the walls and began a violent fire on the Allied column that they saw arriving, which was "imitated by some soldiers dispersed in the camp who, by the natural vivacity of their nation, had taken up arms, doing what was required of them with their fire."[185] So as to not lose any more time, Mölck did not stop, but crossed the bridge in his turn and continued his march. Moving through the French park, it appeared that it would be easy for him to open a passage. He detached a battalion from his column and threw it on the offered prey. Submerged by the wave of assailants, the weak park guard resisted vainly, but as best it could. They did not have the ability to haul off the guns since the teamsters, taken by panic, immobilized the limbers by cutting the traces and fled as quickly as they could to the side of du Chayla's cavalry. The French cannon fell into Allied hands.

During these short minutes, the French infantry approached. The column entered by the right of the camp; "to save the troops some useless steps", the Normandie Regiment had obliqued to the left and "moved by the end of the camp to link that camp to its own camp with

[185] R.A.

a straight line."[186] As a result of this movement, the battalion of the Laval Regiment found itself in the lead, followed by the three battalions of the Crillon Regiment.[187] Seeing the melee around the park, they fired their weapons on the Allies, then, "with bayonet and musket butt", attacked with such vigor that they liberated the French artillery. The Allies attempted in vain to withstand the assault. The linstocks were missing and when some were found, there was barely time to fire some cannon shots.[188] Crillon, arriving behind Laval, chased out the last attackers, killed on the guns all those who did not have time to flee, striking the flank of the Allied column, cutting it into two parts and captured the bridge.

The outcome quickly came. In order to allow their cavalry to arrive and force the passage, the Allied infantry deployed on the right side of the Goutrode Stream, facing Crillon and Laval. At this moment two battalions of the Normandie Regiment, brought forward by Souvré, reinforced the French infantry. The maréchal de camp had rushed back to find this force and had found it, "already laying down its arms and camping."[189]

While, under the fire of the five battalions, the Allies began to hesitate and the French cavalry, pushed forward by du Chayla, put themselves in a position to cross the stream through the woods to turn their adversary, a heavy fusillade suddenly erupted in the rear of Mölck's infantry. Grassin had arrived and joined the battle. Coming out of Massenem, behind the Allies, he had followed them since nothing was there to prevent his movement. Accelerating his march when he heard the sound of the cannon, he arrived just in time to turn the Allied retreat into a rout. As he held the highway, the Anglo-Dutch could not use it and were obliged to flee across the fields or by the bad roads over which, night falling, many chose to use.[190]

During this time, the leading half of the broken Allied column, consisting of 150 hussars, 350 Mölck horsemen, and an Irish battalion that had followed them, suffered an equally bad fate. Mov-

---

[186] *Ibid.* – These details are confirmed by R.A., and Brézé. Besides, they naturally explain the results of the action.
[187]"The Crillon Brigade marched in inversed order, Laval had replaced Crillon in the lead." (Brézé, 12 July)
[188] Brézé, 12 July.
[189] R.A. and R.A'.
[190] "The following day, the local peasants assured me, in truth, of having encountered, a half league away, the advance guard of the army, a number of men who had escaped the battle saying that they did not know where they were going." (R.A.).

ing down the road, the advance guard of these troops collided with Graville. Greatly shaken, he "made many movements, but it was not there where they would have been useful."[191] The hussars seized him. They had barely gained control of their prisoner when a downpour fell on Graville and his captors. Led by their captain, the 50 masters posted at du Chayla's lodging ran to the battle around the park. On the road they encountered Mölck's hussars, and without taking time to count them, they fell on them with such violence, that the Allies, falling into the ditches alongside the road, thought only of their escape. The hussars took their prisoner with them, but at this moment, his horse fell. Graville was saved. "When he lifted himself up, using the branch of a tree to help himself"[192], he found himself alone, his captors had abandoned him. As for the 50 masters, they were already long down the road. "At the risk of being overthrown", they had slammed into Mölck's 350 cavaliers and, before this general had time to become engaged, about a hundred of his cavaliers were cut down. Leaving the leading squadron locked in battle with the French horsemen, Mölck turned to the right and rapidly gained the old road leading to Ghent, along the Escaut.[193] His troops followed him, being fired upon during their passage by the soldiers of the Normandie Regiment occupying the village, the cemetery, and the posts in their camp along the stream. It was a complete rout. Some of the Allies passed in this direction, while others withdrew or attempted to cross the Escaut by swimming to safety. Many drowned and those that succeeded in escaping sought refuge in Bruges and Ostend.

Night had completely fallen, stopping the fusillade that had inflicted considerable casualties on the Allies. Mölck and about 1,500 men could throw themselves into Ghent. Greatly tested, they could take no succor there and they would soon flee before the soldiers of Löwendahl faster than they had fled before those of du Chayla. On the other hand, according to Waldeck, three Dutch battalions and two Dutch squadrons alone succeeded in regaining Alost and Nivone.[194] The Allies had lost 700 men, 1,400 prisoners, two flags, and a standard. The action cost the French 200 men killed or wounded.[195]

[191] R.A'.
[192] R.A. and R.A'.
[193] These details are furnished by the two chroniclers and are confirmed by the statements in favor of Captain de Saint-Sauveur, who led the non-commissioned officers and horsemen of this company. (Work of the King, 1745).
[194] Waldeck, 9-10 July. – Brézé, 12 July.
[195] Crillon to d'Argenson: "I shall present you this time two flags taken by bayonet and

Henry Pichat

Judging that he could not pursue the Allies after such a fatiguing day for his troops and in a land where, with night falling, they risked becoming lost, du Chayla gave the order to encamp on the emplacements prepared by Graville. The following day, he marched out to cover the 6 kilometers that separated him from the Imperial Gate of Ghent, the location of the rendezvous fixed with Löwendahl for the night of the 10th/11th. He brought with him only part of his troops and part of his artillery, leaving the rest to guard the camp at Melle and to erect a bridge at Quartrecht.[196] During the night, du Chayla reached the edge of the covered way of the fortress and posted himself there. He had nothing more to wait for.

While this general dispersed the reinforcements destined for Ghent and Mölck, seeking refuge in Ghent, he served the intentions of Maréchal de Saxe by spreading rumor of the arrival of a considerable corps of French at Melle. Löwendahl began his operation by making some changes to the instruction of 3 July, as a result of the orders daily sent to the expedition. Otherwise, his detachment was weighed down by an important convoy which was bringing the material necessary for erecting bridges over the Escaut in light of the investment of Audenarde. In order that nothing should delay this operation, Löwendahl was to push this convoy as far as Deynse, that is to say, as close as possible to the fortress that he was about to besiege. This method of once again distracted the suspicions that the Allies might conceive for any other enterprise other than the siege of Audenarde. According to R.A., this convoy was at Deynse, on 9 July. According to Brézé, Löwendahl also left Espirres with his troops and his material on the 9th. Godefroy completes this information and permits us to reconstruct the details of the march of Löwendahl's troop. One reads, in effect, in his portfolio, of 10 July: "The troops left this evening. They passed to Courtrai at 7:00 a.m. and arrived at Deynse at nightfall. They departed at 11:00 and, on the 11th, at 2:00 a.m., they attacked Ghent." As the French infantry was unable to cover the 60 kilometers that separated Espierres from this fortress in a single march, it is necessary to conclude that the convoy, the grenadiers, and the partisans left Espierres, on the 9th, very early in the morning. They arrived at Deynse that same day, which merited Méric to put his men in ambush, on the 10th, so that he could conform with the dispositions prescribed by Maréchal de Saxe. Löwendahl, then, set out, marching behind this first column with the

musket but." (Work of the King, *loc. cit.*).
[196] Pons, 11 July.

dragoons, during the night of the 9th/10th, making the movement under the conditions reported by Godefroy and took along, with the passage to Denyse, the Royal Grenadiers, without their baggage. "The troops marched as fast as possible," wrote de Vault; the rapidity of their march, was, in effect, absolutely uncommon for the epoch.[197] At 11:00 p.m., Löwendahl left Deynse, joined en route by the Méric partisans which marched at the head of the advance guard. Behind this latter force came 12 companies commanded by Lieutenant Colonel Morlière and brought along some wagons in which were loaded fascines that had been prepared the day before at Deynse, plans, ladders, and all the tools necessary for the expedition. The four regiments of Royal Grenadiers with d'Hérouville marched next. Chevreuse and the dragoons closed the march.

While everyone marched down the highway, in silence and in the greatest order, convinced that they were going to burn some magazines in Ghent[198], a perfect tranquility reigned over the city. The Austrian governor of Ghent, Kieseghem, had unenthusiastically greeted Mölck and his 1,500 men. They had informed him of the presence of du Chayla, who was a league from Ghent at most; but, covered by the Esccaut on this side, Kieseghem had no serious concerns. Without a doubt, he thought that the Pragmatic Army would send him more reinforcements than the 1,500 fugitives that had entered the place who he appeared to wish to consider more "as voyagers than as defenders."[199] With great difficulty, otherwise, he had consented to grant them lodging in the city. Perhaps, however, he should have mobilized the bourgeoisie militia and to tighten the effective protective flood before the ramparts of Ghent. He did not do anything at all and, when Löwendahl arrived, the major part of the inhabitants and the garrison slept quietly, "having placed, according to the words of a contemporary, all their confidence in some posts which showed themselves less than vigilant."[200]

Arriving at a distance of 3 kilometers from Ghent at about 1:00 a.m., Löwendahl stopped. In order to avoid the confusion that always is to be feared in a night attack, he resolved not to begin the

---

[197] De Vault, p. 51.

[198] "It was necessary that von Löwendahl persuaded his detachment and the others that it was a question of burning the considerable magazines that they would find at Ghent." (Instruction for von Löwendahl, *loc. cit.*).

[199] R.A.

[200] R.A.

action until 2:30 a.m., that is to say a bit before daybreak. There remained only a little time to make the last necessary dispositions, so he hurried. The grenadiers emptied the wagons, loading up with the fascines and the tools and, leaving the highway, turned to the right to close on the fortress across the fields. Löwendahl had resolved to attempt his attack on the ramparts between the Escaut and the Saint-Pierre Gate, because in that location the walls were not reveted and also because the stretch of the shore alone would have been above water if the governor had executed the flooding, as one had had reason to fear. At this moment the lieutenant general informed the troops that they were going to attempt to take the fortress and divided them up for action. Löwendahl proposed to enter Ghent with his dragoons and the Royal Grenadiers, by the Saint-Pierre Gate. As he could not request access from Kieseghem, he wished to wait until his own soldiers, penetrating into the fortress by an escalade, opened the gate for him. To this end, Méric and his men sought to cross the rampart between that gate and the Escaut, while Morlière and his twelve companies reinforced by eight others, attacked the gate itself, covered on the exterior by a demi-lune. At a distance of about 500 meters from this work, the lieutenant colonel had his troops lay down and advanced, accompanied only by two artisan sergeants and four grenadiers. Surrounded by a ditch, the demi-lune communicated with the field by a simple barricade that was closed and guarded. This work linked with the other part of the fortress by a road running across the ditch and leading to Saint-Pierre Gate, over a drawbridge, which was raised at this time, and by a side road that led from the covered road of the rampart. La Morlière sent a grenadier down into the ditch; the water only reached to his belt. A dozen meters wide and six meters deep, this ditch had not been maintained for a long time. The absence of any revetment formed on the bottom of the ditch a thick mud on which much gorse had grown, making it rather easy to cross. La Morlière then called a certain number of men who began to throw their fascines into the ditch, however, carried away by this officer, they threw themselves into the water, climbed up the slope of the demi-lune and reached the crest. They found it covered with a strong palisade, from which, by chance, some of the timbers were missing "forming a hole sufficiently large to allow a single man to pass."[201] This hole discovered, the assailants pushing through it, crying "Vive le Roi, kill, kill!" despite the

---

[201] R.A.

fact that they had been told to keep the deepest silence.[202] Their cries sufficed to put to flight a small post of 20 men that covered the access to the demi-lune from the field. They also attracted the soldiers taking care of Saint-Pierre Gate. They came out in force from their guard detachment, but barely had time to fire their weapons "at the moon"; because Morlière and the 150 men who had followed him hustled them, capturing some, chasing the others off, and entered behind them into the body of the guard post.

The lieutenant colonel then called the rest of his grenadiers, remaining on the glacis, with 50 dragoons, which were attached to them. Led by their officer, these latter crossed the ditch and climbed the slope. The grenadiers followed but arrived with a single lieutenant. The other officers, too old to perform such physical feats, had remained on the glacis,[203] then they set about breaking down the drawbridge by cutting the changes. It eventually fell. Morlière's troops then broke open the access gate to the demi-lune, aided, from the exterior, by Löwendahl's impatient dragoons. During this time, Méric had executed his own escalade, without encountering any obstacles, and his partisans mixed with Morlière's soldiers vainly attempting to break the interior hinges of the Saint-Pierre Gate with great blows of the butts of their muskets. "One would have broken all of them in the army," said a chronicler, when an officer came up who recommended searching for the key or a blacksmith. The first man that they awakened came, prepared his tools, with which the hinges were broken, and the French troops entered.[204]

Day broke and it was necessary not to give the defenders time to organize themselves. The bourgeois militia showed no desire to become involved, but some pillaging had begun. Löwendahl withdrew the greater part of his troops to the glacis, towards the citadel, as much as to stop the excesses of the French soldiers, as to "avoid animating the very seditious population of Ghent." [205] Otherwise, since the entry into the city, the general and his dragoons moved rapidly on the citadel, encountering no resistance. The little diligence shown was when

---

[202] *Ibid.*

[203] "It is appropriate to observe that in the formation of the Royal Grenadier regiments, they were given veteran officers and preference was given to those who had retired into the provinces with pensions, but who, because of an advanced age, found themselves unable to cross the ditch and climb a rampart whose slope was very steep." (D'Espagnac, *loc. cit.,* Vol. II, p. 104 n.).

[204] R.A.

[205] R.A.

Kieseghem sought refuge with an English free company before the French arrived.

As for Mölck and his 1,500 men, they had fled out the Ecluse [Sluice] Gate, at the first shot, without waiting for the French guards to capture the gates. It would be necessary to besiege Kieseghem. However, this operation did not appear that it would be either long or difficult, but it could have been avoided if one believes a contemporary who reproaches Löwendahl for not having sufficiently prepared his surprise attack. "Some dragoon commanders have assured me that they were in the city for more than an hour at the head of their troops without receiving the least order to act" and, according to the same author, Mölck's flight furnishes the best proof of this assertion.[206]

Be that as it may, while Löwendahl moved against the citadel, d'Hérouville and the Royal Grenadiers moved against the Imperial Gate before the shocked eyes of the Ghent bourgeois, surprised to find these new guests in their city, and have perceived nothing "because of the lack of resistance."[207] After the alarm given by Mölck, the intention of the Allied garrison was to move to the side of Melle. They would later discover du Chayla and fired a few ineffective shots at him. Unable to think of launching an assault over the rampart covered by the Escaut, the French general had contented himself with responding and waiting. He finished by finding, no doubt, that he had not waited long enough, because he had begun to return to Melle at the moment when d'Hérouville's grenadiers opened the Imperial Gate with cries of "Vive le Roi!" Du Chayla returned; "everyone had lunch in Ghent; it was too late to sleep."[208]

There only remained to capture the citadel where Kieseghem lacked provisions. One could foresee that the defense would not be more vigorous than that of the city itself. The governor, summoned to surrender, refused, "wishing at least to see some cannon."[209] The French prepared to show them to him.

The following day, 12 July, bringing from Deynse with some other materials, the engineers made their dispositions; because even though they had to rapidly execute the pretense of a siege, they did not wish to make any errors in the rules of the art. These exaggerated cares gave the English free company the time to flee down the Den-

---

[206] R.A. No document justifies this.
[207] R.A.
[208] R.A'.
[209] R.A.

dermonde. It moved so rapidly because it contained a large number of French deserters, who little desired to encounter the Grand Provost of the French Army. To keep Kieseghem from fleeing himself, the camp at Melle was moved on the 14th, to Destelberge, on the left of the Escaut.[210] Finally, on the 15th, at 6:00 a.m., the first parallel, pushed to the flying sap, was completed. [211] Six cannon with four mortars went into action when Kieseghem raised the white flag. He asked, with the honors of war, authorization to send four covered wagons to the Allied camp. Löwendahl understood that the military chest of Ghent was in the citadel. The request of the governor was rejected. In addition, the attitude shown by Kieseghem did not authorize him to make any demands. He was to capitulate with his garrison as prisoners of war.

The French captured, at Ghent, 500 prisoners and a large number of horses; 500,000 rations of forage; 11 new cannon recently shipped from England to replace the heavy artillery lost at Fontenoy by the English; 150 milliers of powder and large quantities of ammunition. The French found, in the arsenal, 15,000 sacks of oats and wheat, and everything necessary to equip 8,000 men.[212] Only the food, artillery and ammunition was kept, the rest was sold publically. The total value of what was captured was 10,000,000 livres.[213]

Du Chayla was named governor of Ghent, and the city was provisionally occupied with two regiments of dragoons and the Royal Grenadiers. Lieutenant General, the Prince de Pons, took command of the troops commanded at the Destelberge camp.[214] On the vague word that only 800 men defended Bruges, the Prince de Pons, sent four companies of grenadiers, 200 fusiliers, and 150 Grassins to capture that city. They left Destelberge during the night of the 17th/18th, under Souvré, "who was to act on what he learned about the garrison and to summon it, if it was found to be as weak as indicated."[215] When the advance guard, formed of the Grassins, presented itself before Bruges, it encountered no resistance. The magistrates only declared that they were ready to open the gates "to a principal officer." Immediately as Souvré arrived, the French entered Bruges. This city had contained only a few gunners for its defense; but, since the capture of Ghent,

---

[210] Brézé, 12 July.

[211] D'Oyré, commander of a brigade of engineers, to d'Argenson, Ghent, 15 July (Corresp. A.F.).

[212] R.A. – R. A'. – Brézé, 12 July.

[213] Löwendahl, *Mémoires*; Brunet, *Tableau de la guerre de 1740-1748.*

[214] Pons, 18 July.

[215] Pons, 18 July.

the Allies sent the materials stored in Bruges to Ostend, the fate of this latter place being very precarious.[216] The Grassins rapidly pushed through Bruges to the Damme Fort; but they only captured its governor, the garrison having fled to Ecluse.

These captures assured the French the establishment of a position on the left of the Escaut. As a result, the Prince de Pons moved to the Melle camp, on 19 July, to guard the bridge at Quartrecht and the following day Contades occupied Bruges with [the] Carillon [Regiment?] and a regiment of dragoons detached from Ghent.

The French prepared to draw contributions over all the land to the north of Ghent and Bruges, as far as the Dutch Fortresses of Ecluse, Sas de Ghent, Axel, Hulst, and as His Most Catholic Majesty was not at war with the Republic, the Maréchal de Saxe that the deputies of Their High Powers to mark, in a clear and appropriate manner, the territory of their states so that the French troops would not make requisitions on the territories belonging to the Queen of Hungary.[217] The purpose of this was to skilfully attract the attention of the Dutch of the French Army's proximity to their frontiers. Despite everything, the Provinces "thought themselves assured by flooding their own lands to save the part that was threatened by us [the French]."[218]

The populations did not accept this measure with the same enthusiasm as in the previous century. The Dutch commander of Ecluse sent an engineer and some soldiers to open the dikes on the Bruges Canal. The peasants killed the first and turned the soldiers over to du Chayla.[219] From the following day 200 Grassins guarded the canal.

The capture of Ghent was so successful that Löwendahl was quickly assigned to undertake the siege of Audenarde and it appeared that this officer would easily complete the operation, whose success was not in doubt.[220] About 1,500 men defended this place under Austrian Major General Mac Hugo.[221] The fortification was well main-

[216]R.A. – Bruges was defended only by a simple wall.
[217] Maruice de Saxe to the Deputies of Their High Powers, the States General, Borst, 14 July (Corresp. A.F.).
[218]R.A.
[219] *Ibid.*
[220]Contemporaries of Löwendahl were not of this opinion; R.A. wrote, "All the siege works of Audenard were as bad as they were quick. This general wished to distinguish himself in the eyes of his master by the great activity that he showed, but which was unskillful in the eyes of enlightened men."
[221]A state of the garrisons of the fortresses in the Lowlands besieged by the French in 1745, held in the archives of La Haye, states that the garrison of Audenarde had an Austrian battalion (Baisrugg), an English regiment (Beauclerc), and a Dutch Regi-

tained and powerful. By virtue of its position on the Escaut, Audenarde could make effective movements of the water because of the importance of this river and the force of its current.

The convoy sent, on 9 July, to Deynse, under the conditions previously explained, took the road from Audenarde on the 11th.[222] The following day the French troops erected a bridge at Eynem and another at Heurne, both of which were upriver from the fortress quickly invested by the Fitz-James Cavalry on the left bank and two regiments of dragoons on the other bank of the Escaut.[223] These latter troops sent from Ghent established themselves on Mount Cerisier, which dominated the fortress from the right bank. During this same time, the artillery train prepared at Tournai moved by the river.[224] Löwendahl was to reinforce this material with the 11 English cannon captured at Ghent as soon as the reduction of the Ghent citadel was complete.[225] On 14 July, three brigades of infantry arrived from the Borst camp and relieved the dragoons that surrounded Ghent, bringing the infantry of the siege corps to 18 battalions.[226] The army was to furnish, in addition, fascines and workers.[227]

Löwendahl attacked the fortress by the right bank of the Escaut. He supported his front on the left of the village of Leupeghem and extended his lines to the river thus embracing the three bastions of the southwest front of the defense. From the heights of the Cerisier, the French cannon could strike the least recesses of the city. The French infantry camped between this hill and the Tournai highway and it appeared that the French soldiers suffered much by their closeness to the city.[228] Thirty-two hundred workers and the fascines had arrived from Borst and a trench was opened at 100 toises from the covered

---

ment (Brawn).

[222] Brézé, 19 July.

[223] Fitz-James came from Tornai as the escort for the convoy going to Deynse.

[224] Instruction for Count von Löwendahl, *loc. cit.*

[225] "I shall send, tomorrow, artillery engineers so that they can arrive without losing time before Audenarde. I plan on being there tomorrow." [Löwendahl to Maruce de Saxe, Ghent, 15 July (Corresp. A.F.)].

[226] The Picardie, Bouzols, Seedorf, and Monin Regiments totaled 16 battalions, plus two battalions of the Löwendahl Regiment.

[227] Brézé, 19 July.

[228] "The lives of the soldiers were of so little consequence during the siege that the Picardie and Bouzols Brigades, camped between the Cerisier and the fortress, were so close that the shot, after having crossed the camp, passed almost as far beyond it, killing many in this manner." (R.A.)

road, during the night of 18/19 July.[229] In the morning, the first parallel was 110 toises long and part of the French guns were emplaced and had begun fire. The construction of the batteries was barely completed. The lack of protection soon became the source of new casualties.[230] There were two batteries; to the right (6 mortars) and to the left (8 mortars and 5 cannon) of the parallel. There were also here others, containing a total of 30 cannon, on the crest of Cerisier. A fourth was prepared on the mountain on the extreme left of the parallel to direct its fire on the Tournai Gate.[231]

From the opening barrage, the French projectiles set the city on fire. The Allies' repost permitted the French to observe that they only had a few, small caliber guns. However, they caused sufficiently serious numbers of casualties among the French. Finally, during the night of 20/21 July, the French sap closed on the forward ditch of the bastion before the Brussels Gate, the center of the defensive front. The French workers immediately began the construction of a bridge made of fascines to cross the ditch. At daybreak, all the French artillery began firing, redoubling the intensity of the fires in the city and prepared the way for the infantry's attack. The governor, however, raised the white flag before the attack began.

The firing quickly ended and hostages were exchanged. However, Mac Hugo did not complete his proposals until the 22nd, at 5:00 a.m. They were unacceptable, because the Governor of Audenarde asked only "to send two officers to von Königsegg to receive his orders on what steps to take."[232] Löwendahl threatened to resume fire and as the construction of the fascine bridge was pursued throughout the day of the 21st, the governor capitulated. He did not doubt that his ramparts would fall to the assault. The garrison could leave the fortress with the honors of war, but they would lay down their arms at the city's gate.[233] The French had lost 100 dead and wounded and, in such conditions of rapidity, "one could not have taken the fortress at a better price."[234] Sixty hours from the opening of the first trench was sufficient to reduce its defense. Mac Hugo had not taken advantage

[229] Brézé, 19 July.
[230] "Löwendahl had only given 24 hours for the construction of the batteries and appeared unhappy when he was asked to give twice that time, which was the time ordinarily required to build them solidly. He was obeyed." (R.A.).
[231] Brézé, 19 July.
[232] Mac Hugo to Löwendahl, Audenarde, 22 July (Corresp. A.F.).
[233] Audenarde Capitulation (Corresp. A.F.).
[234] Brézé, 19 July.

of any of the defensive advantages offered him by the Escaut. "This governor," wrote R.A., "was so little disposed to sacrifice himself, that the third day the trench was open, not having more than a few fascines on the edge of the forward ditch, where they wished to erect a bridge, he surrendered. He still had two ditches filled with water and all the forward ditch was free, because that portion of the bridge that was constructed during the course of negotiations for the capitulation, it was not more than two–thirds of the way across the ditch. "If the enemy had shot [at us], [while we were] building that bridge, which would have cost us at least 300 men every time [as it had been] placed in an open [unprotected] location; it [the bridge] had been the work of 36 hours and we would have had to start all over again as long as such had been the wish of the governor [to continue resisting]. Two days later I attended an experiment that had been made in order to break [destroy] it, against which it did not resist and it is impossible in that location for a bridge of that type to resist all the surges of water [tidal surges] that it would suffer every day."[235]

On 13 July, on the order of Maréchal de Saxe, Clermont-Gallerande left Binch to go to Chièves. This lieutenant general no longer had any reason to remain in his first position and the elongation of the army left him somewhat hanging in the air.[236] He learned, upon arriving at Chières, that the garrison of Ath contained only 1,400 men; because two battalions had left the fortress the day before.[237]

The Allies, however, were not unaware of the march of this detachment which, now, appeared to be preparing an investment of Ath. They did not appear to fear this movement; recent events having given them more concerns, fears even, that were much more important. During the morning of 10 July, in effect, the Allies having learned of the check suffered by Mölck and the retreat of the two battalions to Alost and Nivone,[238] reiterated the order to these troops to move anew against Ghent, but this time, they were to pass by Dender-

---

[235] R.A.

[236] According to Brézé, Clermont-Gallerande affected his movement in two columns. The three battalions forming the right of the march, and following the roads left out, crossed the Haine at Trivière, later passing through Strépy, Ville-sur-Haine, Saint-Denis, Heriches, and Lens. The cavalry, forming the left wing, crossed the Péronne Stream at a ford near Binch and reached its camp by Villers-Saint-Ghislain, Havré, and the Casteau moors between Saint-Denis and Lens. These latter troops marched across country.

[237] Clermont-Gallerande to Maurice de Saxe, Chièrres, 14 July (Corresp. A.F.).

[238] Waldeck, 10 July

monde and follow the left bank of the Escaut. At 10:00 a.m., that same morning, the Pragmatic Army broke camp, formed in two columns and marching to its right, it moved to establish itself between Nivone and Grammont, occupying the right bank of the Dender, sometimes in one line and sometimes in two, according to the dictates of the ground. Two battalions and six guns remained in Grammont, where six battalions and a Dutch cavalry regiment occupied the heights, no doubt because, during the march, the Allied spies had reported that the French were preparing to attack. It appeared that the Pragmatic Army sought to move closer to Ghent. The movement executed, on the 10th, as well as the orders sent to the debris of Mölck's battalions, can be attributed to this intention. The full scope of these movements also indicates a growing awareness of the situation, because during the march, the Allies learned that Löwendahl was moving on Ghent. At 3:00 p.m., on the 11th, the Allies received solid word that Ghent had fallen to the French.[239] They then recalled the two battalions sent from Alost and Ninove, which were running a great risk of being swept up by Löwendahl upon their arrival and the Allies prepared to take up a post between Alost and Assche.[240] Königsegg sent General Saint-Clair, Quartermaster General of the Pragmatic Army, to prepare the camp, fix the departure of the columns for 7:00 p.m., and to send the movement order to Waldeck.[241] The Prince barely took the time to read the note from the Marshal and ran to see the latter, who he found with Cumberland.[242] Without completely reporting the deliberations of the council of war held with Königsegg; Waldeck recounted only that it had been resolved to march on Brussels that very night.[243]

The artillery and the baggage of the right wing were to put themselves in movement at midnight. Profiting from the overtures prepared for the departure on Alost, this convoy reached the highway

---

[239]Waldeck, 11 July.

[240]*Ibid.*

[241] "The English cavalry passed over the Nivone Bridge; the Dutch followed the English. The infantry marched at 8:00 p.m., in two columns by the openings made beyond the Dendre. The artillery and the baggage formed a third column on the right of the second." (Waldeck, 10-11 July.)

[242] "Upon [receiving] this note I quickly went to the Marshal." (Waldeck, 10-11 July).

[243] Waldeck, 11 July – If the Prince did not give the reasons for this retreat, a contemporary gives them clearly: "The Allies, weakened by so many losses, retired through Brussels and entrenched themselves on the canal, in a camp that de Vendôme had previously called the 'camp des poltrons' [coward's camp], because one could not be attacked there." (*Historie de la guerre de la Succession de l'empereur Charles VII*, attributed to du Muy).

from Ghent to Assche and followed it as far as the Kockelberg heights. The infantry of the second line, as well as the cavalry of this wing, leaving at 2:00 a.m., moved along the same road. As for the first line, it reached, in the same formation, the highway from Ghent to Brussels, by Moerbeck, and camped at Kockelberg with the troops that had preceded it. On the left wing, also, the baggage and the artillery marched in the head by Denderwindicke and Neygène on Anderlecht. All of the troops of this wing united into a single column, the infantry in two lines in the lead, the cavalry in the rear. The forces remaining at Grammont served as a general rearguard. The quartermasters, finally, led the infantry so as to move rapidly forward to mark the camp as soon as the order was given.[244]

Despite the simplicity of these dispositions, the march was very difficult. The departure of the first elements did not occur until 2:00 a.m., the artillery and the baggage, no doubt, having suffered some difficulties assembling in the deep darkness. Because the convoys were imprudently placed in the head of the column, the movement lasted until 5:00 p.m., on the 12th. The Pragmatic Army had lightened its march by sending its baggage to Brussels several days earlier. At the price of a 15 hour march, the left wing, commanded by Waldeck, finally reached Anderlecht. On the right, Cumberland's cavalry had reached Kockelberg, but the Duke's infantry had not passed Assch, where it encamped.[245]

The Pragmatic Army could not remain in this situation. Königsegg decided to put himself in defense behind the Brussels Canal as soon as possible. As of the 13th, the English infantry had rendezvoused with the Allied, and, on the 14th, the Allies resumed their march.[246] That same day, finally, Cumberland and the right wing encamped at Dieghem and the following day, Waldeck and the left wing, marched in two columns, rejoining the Duke after a march of two hours.[247] Camping at Vilvorden and Schaerbeck, under Brussels, the Allies were organized in a position that did not appreciably change until the end of the campaign.

---

[244] Waldeck, 11-12 July.
[245] Waldeck, 11-12 July.
[246] *Ibid.* 13 July. – On the 13th, also, the detachments from Enghien and Alost rejoined the Pragmatic Army.
[247] *Ibid.* 14-15 July. – The passage through Brussels alone required the division of the left wing into two columns.

Twenty days had sufficed to considerably reduce the maneuvering room left open to the Pragmatic Army. The Maréchal de Saxe, master of the Upper Escaut, holding at his mercy the fortresses along the Dender, left by the Allies almost without defense, such as Nieuport and Ostend, which were completely isolated. It had become unnecessary to take these last two fortresses, they had become useless to the English and the direct communications between the London Court with the Pragmatic Army were now only possible through Holland. No one could envisage in which direction Maréchal de Saxe was going to direct his next blows. The Allies had suffered significant material losses, without being able to resolve the indecision which Maurice de Saxe had so carefully cultivated in them. If, in effect, the Allies were resigned to the loss of the important fortresses along the Dender, the security of Mons and Antwerp caused them, in contrast, great alarm. The retreat, though it may appear a bit premature, of the Pragmatic Army to the right bank of the canal, had been inspired only by the need to defend these two places. However, would Maurice attack Mons first, with an eye towards clearing the French frontier, or would the French cannon, ruining the ramparts of Antwerp, tear this port from Holland?

Finally, during these days of bad fortune, the state of morale in the Allied Army appeared to become most precarious. "Cumberland," wrote the spy Aulent, "had assembled the principal officers of the army, proposing them to mix the three nations together and to put an English regiment with a Dutch and Hanoverian Regiment, and to thus mix all the army. He hoped that this would promise them the greatest success. The Dutch did not wish to accept this proposition, which produced so much confusion among them that without Königsegg's great prudence the Anglo-Hanoverians would have fired on the Dutch, saying that they were not reliable and that it was them that one wished to sacrifice. The proposal was sent to the States General and they awaited the result. If the States did not send a favorable response, it could well produce a change in the project as well as a separation between the Allied troops."[248] This tardy proposal made by Cumberland was not adopted. However, it would not have made any improvement in the state of morale in an army where everyone complained of being deceived or a victim.[249] Aulent collected all the

---

[248]Aulent to d'Argenson, Philippeville, 13 July (Corresp. A.F.).

[249] "There came to me two reliable men from Ath who ensured me that there would be in this city only 550 Dutch and 770 English, the two troops detached of the army,

remarks which were exchanged among this multitude; "they are ready to fire on one another, they say, and I do not believe that the campaign will end without this happening."[250]

In the fortresses distrust reigned as great as it was in the army; but the governors made preparations for the defense that the inhabitants did not appear to see with much enthusiasm.[251] The excellent weather promised an abundant harvest; the fear of ruin became insupportable to the country folk, who, with those of the cities, thought that in lieu of resisting the French, it was preferable to oblige the governors of the Queen to surrender their fortresses.[252] Maria Theresa had, otherwise, posted on 11 July, in the County of Namur, that she enjoined immediately bringing forward the grants accorded for the rest of the year.[253] The actions previously taken by the States of the Provinces of Ghent and Bruges had inspired this action, which was otherwise very badly accepted.[254] The magistrates of these cities had, in effect, preferred to treat directly over contributions with Séchelles, Intendant General of the French Army, without awaiting the arrival of Maréchal de Saxe's requisitioners, and, if such became the general practice, the provisioning of the Allies and the output of the tax was likely to decrease greatly.[255]

In sum, the events of July had further accentuated the reciprocal situations created by the victory of 11 May.

Upon his departure from Leuze, on 4 July, Löwendahl alone held the Marshal's instructions. He wrote 12 years later "that one could not pay enough attention to spies" and who, joining example to precept, himself maintained numerous spies in the Allies' army, in the various fortresses, and in the Council of the States of Brussels, could not leave the secrets of his projects to the mercy of indiscrete chatter.[256] Nothing transpired, and the coup de main against Ghent could be delayed by 48 hours without the surprise being less great, both with

---

which reciprocally complain of being sacrificed in a place where they could not defend themselves while the army is on the other side of Antwerp." (Clermont-Gallerande to d'Argenson, Camp at Chièvres, 16 July (Corresp. A.F.)].

[250] Aulent to d'Argenson, Philippeville, 16 July (Corresp. A.F.).

[251] Aulent and Thirion to d'Argenson, July, (Corresp. A.F.).

[252]*Ibid.*

[253] Aulent to d'Argenson, Philippeville, 13 July (Corresp. A.F.).

[254] "This produced some murmurs." (*Ibid.*)

[255] Maurice de Saxe to Folard, Borst, 11 July (Corresp. A.F.).

[256]"It is necessary to send spies everywhere, with the generals, with the officers, and even with the sutlers." (Maurice de Saxe, *Mes Réveries,* p. 134.)

the Allies and with the French Army. Also, when the King awoke, on 11 July, according to the convention established earlier between the King and the Marshal, the Marquis de Sourdis, aide-de-camp to Maurice, presented himself to the King with a sealed basket, from which he drew a loin of veal and which, Louis XV, upon seeing it, announced to his assistants the capture of Ghent, the surprise that painted the faces surrounding the King were certainly not false.[257] It is undoubtedly under the impact of his emotion, following this scene that an anonymous friend of Count d' Estrées wrote: "Everyone was misled there and I must render justice to the Maréchal de Saxe that his project was beautiful, extremely secret, and extremely fortunate."[258]

The general enthusiasm rose as a result of the operations of July, the brilliance of which lacked nothing, leaving little room for sentiments other than admiration for the Marshal who had conceived them. The general charged with the execution of his orders were not sparing in his criticisms. "If the reports were intended to be read in public, one would shout against the affected omission of all the wise precautions which the success of events supposed that the general did not fail to take; but as I write only for myself, I do nothing but express the pure and simple truth," said a friend of Count d'Estrées.[259] After this edifying preamble, he pitilessly enumerates all the omissions made by Löwendahl in the preparation and the execution of his coup de main." He adds that "the books are in this instance of an infinite utility."[260] He concludes that if Löwendahl "had taken a look," he would not have exposed himself to the failure resulting from being surprised and compromising the enterprise in which Löwendahl omitted so many essential precautions.

After this indictment it would be unjust not to present the counter-point.

"These enterprises," wrote Löwendahl, "conducted with as much celerity as success, had a contrary effect on a great number of French generals. They excited their jealousies to such a point that they said that the King was affronting the sensitivities of the French nation in employing two German generals on this most important of expeditions, as if there was not, in the French nation, anyone that had their

---

[257] Ghent was known for the excellence of the veal that was raised there. (D'Espagnac, *loc. cit.*, p. 106).

[258] R.A.

[259] *Ibid.*

[260] *Ibid.*

capacity. Everyone should be authorized to say that His Majesty owed his conquests to a couple of Germans. Otherwise, they said that he it was fortunate for them that the King had participated in the campaign personally, because, without that, they would have found less subordination and obedience in the national officers."[261] It remains to d'Argenson to decide between these opinions. This Minister of Foreign Affairs has, however, left generally unflattering portraits of Maurice de Saxe and Löwendahl; however, he also writes; "we have never possessed fewer officers suitable as generals. The needs of our affairs has reduced us to engage foreigners; there are vain complaints and to murmur from above, their works and their success, without discourse, are worth more than the eloquence and the promises of the others."[262]

On 9 July, it was learned, in Borst, around 11:00 a.m., of the encounter at Westrem.[263] At first it was seen as an episode without consequence, where the ardor of the excellent Grassins was so often the talk of the army. However, when during the evening one knew of the action at Massenem, fears vanished. Some have suggested to Maurice the idea of sending reinforcements. Not sharing the general excitement, the Marshal refused. In the middle of the night a messenger brought news of the capture of du Chayla's artillery. This raised some concerns. The Marshal was awakened: "I am certain," he said, "that the detachment of the enemy consists of no more than 5,000 to 6,000 men. What you tell me about the artillery is impossible. If the enemy has encountered du Chayla, he is in the doldrums. Du Chayla has defeated them, and I shall sleep peacefully."[264] The narrator of this story saw here only "a trait of the Marshal that clearly makes known the calmness of his character." A friend of the Count d'Estrées contented himself with saying that one took few precautions and concluded philosophically: "the general, upon succeeding, is approved: it is a certain law of war."

Du Chayla and his assistants were not forgotten. "I do not wish to speak to you about our generals," wrote an actor in the episode at Melle, "but you will probably find them bad. It it is pointless to speak of du Chayla. As for de Souvré, at the first rumor that the cannons had been taken, he thought he had lost his baggage. He said nothing else to all those he encountered, adding that he had lost 10,000

---

[261]*Mémoires* by Löwendahl, *loc. cit.*
[262] Rathery, *loc. cit.* Vol. IV, p. 207 and following.
[263] R.A.
[264] R.A..

écus. De Laval and Crillon did not see it as being so critical when everything had calmed down. There was even between them an unfortunate scene, the first wanting to attack the enemy, the other wishing to take up a defensive; it had been said that he still feared the enemy. De Graville was wounded, but I doubt he knew by whom. If it is permitted for me to relate my judgment, there was never a more favorable occasion for du Chayla to merit, on the part of a good general officer, to become a marshal of France, or to let it be known that he had the necessary talents to rise to that dignity."[265]

Such are the impressions of the chroniclers that show what one of them called "the back side of the medal." As for the Allies, they collected only their scorn. "With equal defenders of these fortresses," one said, "we could be assured of succeeding at everything," and another, with regards to Kieseghem, added: "If this man had won, we would have barely saved him the appearances [of a victory] by the few precautions that we had taken against him; if he had not, which I believe, we had a quite bad opinion of him."

The battle of Melle inspired one of the critics with the most judicious joke, "here is roughly the affair; if it is necessary to credit a miracle, it will be in the honor of Providence. Our generals could not claim it, and if, finally, the good God does not wish to take credit for this event, it is necessary to praise de Crillon, de Laval, and also the good will of our brave infantry."[266] [The joke is that God is not taking credit for the victory.] The result of this campaign would make even more violent the contrast between the valor of the troops and the weakness of the organs of the senior command.

[265] *Ibid.*
[266] *Ibid.*

CHAPTER III
The Results of Operations on the Escaut.

The order given, from the end of May to Prince de Conti, to not present any obstacle to the Imperial election, did not simplify the mission of the French army in Germany. De Traun, replacing Maréchal de Bathianyi at the head of the Austrians coming from Bavaria, sought to join d'Aremberg. "Their project is well uncovered," affirms Conti, who estimated that this maneuver would give the Allies a superiority of around 10,000 men.[267]

Louis Charles César Le Tellier by *Louis-Michel van Loo, 1759*

The last instructions that the Court sent Conti were to unite his forces between the Main and the Necker. Foreseeing that his situation could become difficult, Conti had sought to prevent the junction of the Queen of Hungary's two lieutenants by throwing himself against von Traun, who was closest to him. As von Traun had moved his line of march past d'Aremberg far to the east of the Neckar, the project of awaiting him became unrealizable. However, it appeared that Conti rather easily played his role in the junction which he could not oppose and which another possibility, more serious according to him, was more worrisome. If, in effect, d'Aremberg left Bonn, moving up the Rhine to Mainz, to cross the river on the territory of the Elector, each day would be more favorable to the cause of Maria Theresa, while Conti feared that he would be obliged to defend

---

[267]Conti to d'Argenson, Pflugstadt camp, 5 June (Corresp. A. A.).

his own, now threatened, communications. He had to move closer to the Rhine or move to the left bank the forces necessary to contain d'Aremberg. Could he respond [to the King] that he was not forced to move there with all his army to cover the French frontier?[268] How does one conciliate this necessity with the formal orders that he had received to remain on the right bank of the river?

At this moment, besides, Conti commanded no more than 48,000 men organized in 114 squadrons and 81 battalions, of which 7, among the latter, were employed as guards for the bridges and magazines on the lower Neckar.[269] One month and a half later, events would show that Conti's fears were not chimerical.

The Prince hoped for a moment that his situation would be improved because of the victory won by Frederick II over the Austrians at Hohenfriedberg. Informing Conti of this happy success, the Prussian Monarch profited from the circumstance to express his ideas to the French general and "that he desired that Conti act to second them."[270] Very preoccupied, the Prince was not disposed to extend a willing ear to Frederick's plans. His own instructions were too precise "for him to consider any contrary consideration." Conti hoped, as a result of this victory, that he would receive new orders.[271] And those orders did come.

The Minister sent new instructions from which it appeared that the old plans had not been abandoned and that the new success of the King of Prussia raised the hope of delaying the Imperial election. However, according to Conti's own words "we will not fulfill much of our projects in the Empire, having recrossed the Main." The French could no longer impose on the Elector of Mainz or on the Circles of Swabia as they had retired beyond the Necker. Finally, as Louis XV had declared that he did not wish to place any obstacle in the path of the Imperial election, the Court appeared to have decided to manifest

---

[268] It could be perceived that not believing the in election or the declaration of the Circles, the Allies believed it necessary to cross the Main to unite beyond it, either to fight the French army, or solely to intimidate it and oblige it to move back on the left bank of the Rhine. In these cases, the advantage of defending either the defiles or the river, although it was fordable, would compensate for the numerical inferiority of the French army. They could dispute the passage to the Allies and by virtue of the choice of his strong position, perhaps engage them in a successful battle. (Conti to d'Argenson, Dieburg Camp, 7 June (Corresp. A. A.)].

[269] Situation of the Army of Germany, 11 June (Corresp. A.A.).

[270] Conti to d'Argenson, Dieburg Camp, 10 June (Corresp. A.A.).

[271] Conti to d'Argenson, Dieburg Camp, 10 June (Corresp. A.A.).

a certain offensive.[272] It was too late to feed this hope and it would have been preferable to have openly informed Conti that he should oppose a diversion on the left of the Rhine. Such a movement would oblige the Army of Flanders to send detachments into Lorraine or Alsace and thus Maurice de Saxe would lose the numerical superiority that he had acquired. The news sent by French agents became worse and worse.[273] Traun and Bathianyi, who came to replace d'Aremberg, appeared to wish to unite on the Nidda.[274] On 23 and 24 June, information became more precise. Traun had crossed the Main at Lohr and Bathianyi was at Budingen. One foresaw that their junction would occur near Gelenhausen.[275] Beginning, in reality on 28 June, between Vechtersbach and Orbe, the operation was completed on the 30th, assembling 60,000 men before Conti, who, happily was on his guard, and was able to assure his safety by moving closer to the Main, whose fords had become impassable because of the constant rain. Finally, on 4 July, Ferdinand of Lorraine, spouse of the Queen of Hungary, the Imperial candidate, took command of the Allied Army.

It appeared that the soldiers had nothing more to do than to lead Ferdinand in triumph to the foot of the throne of the King of the Romans; because, at that time, one had vainly sought to constrain the influence exercised by Austria over the Electoral Corps.[276] In effect,

[272]"If with your reunited forces, Your Highness judges it proper to cross the Main to support or seek combat with our enemies, according to what you judge possible or appropriate, His Majesty gives you full liberty. It is certain that our army, on the left of the Main, influences on the Elector of Mainz and the Circle of Franconia no more than if it was on the right of this river. However, with regards to the Circle of Swabia, it is no less open by report to us than when we were on the left of the Main and the enemy on the right is no longer any support to it. It would be different if Your Highness thinks himself obliged to recross the Rhine. This is the principal reason that His Majesty desires that Your Highness holds himself on the right of this river. His resolution does not vary on the resolution that He has taken to not bother the liberty of the election, but to the contrary, he wishes that it not be troubled by his enemies." [D'Argenson to Conti, camp below Tournai, 12 June (Corresp. A.A.)].

[273] Renaud to d'Argenson, Coblentz, 12 June. – Beaumès to d'Argenson, Bonn, 14 June (Corresp. A.A.).

[274] Conti to d'Argenson, Dieburg Camp, 15 June (*Ibid.*)

[275] *Ibid.*

[276] "It was not the enemy army that delayed the election, the pact was completed, the majority of the electors had won. They wanted the Grand Duke to be the Emperor; it would only take force to oblige them to take the decision. It was certain that military successes had some influence, but it is also certain that it would be to flatter oneself to imagine that only they could change principles of enemy courts whose ministers and the masters themselves are bought and sold by ours. It would have been desirable that we be able to go to their markets; it would be still more desirable, but costly, to seek to

on 16 July, the Grand Duke Ferdinand entered Mainz to the sound of acclamations and salvos of artillery, after having thrown a corps of light troops, under General Bernklau, to the right bank of the Rhine. Finally, on 19 July, Conti found himself, in his turn, obliged to cross this river, "in order to avoid" he said, "the disadvantages whose results could be most regrettable for the army and the King's frontier."[277]

A courier of His Serene Highness brought this bad news to Ghent, on 26 July: "I am greatly affected," wrote the Prince, "by the pain that I feel that Your Majesty will have for the movement that I have been obliged to execute." Louis XV had received this missive during the course of the triumphant trip that he pursued in his new conquests. It did not appear that the King expressed many regrets on the fate of affairs in Germany, if one is to judge by the response of the Minister to Conti.[278] Vainly the latter wqas assured that he was examining that which it was possible to do, that he would move to penetrate the Allies' designs, to make new dispositions, to reflect and to plan. Vainly France's Ally the Elector of the Palatine, uncovered by the retreat of Conti, presented his complaints, enumerated his fears, and requested that if one would not second him, "that one must leave him at least the means to take appropriate measures; the essential thing, according to d'Argenson, was that the Army of the King was whole; that it occupied a good and secure position and that finally our magazines and frontiers had nothing to fear."[279]

Conti's army, established on the left bank of the Rhine, was not to move, so long as there was no more Allies on that side of the river.

"The Prince has just played the Gille[280] on the banks of the Rhine," said Frederick II, irritated by the retreat of the French Army

break them by the same means by which they were purchased." [Conti to d'Argenson, Steinheim Camp, 5 July (Corresp. A.A.)].

[277] Conti to the King, Hochheim camp, 21 July (Ibid.). – Bernklau was moving to recross to the right bank of the Rhine almost immediately.

[278]"That which can cause concerns in the position of Your Serene Highness is less interesting in its military considerations than its political considerations and Your Serene Highness knows that I have thought, for a long time, that the delay in the election of the Grand Duke depended less on the presence of an army on the territory of Frankfurt than the claims of a competitor who has not yet declared, and that the division of interests between the Princes of the Empire is more capable of giving birth to than the force of arms."

[279] Conti d'Argenson, Orcheim camp, 25 July (*Ibid.*) – Elector of Palatine to Conti, 19 July (*Ibid.*) – D'Argenson to Conti, Ghent, 27 July (*Ibid.*)

[280] Translator: The "Gille" are giants in carnival parades in Belgium.

of Germany.[281] Was this malicious assessment well founded? It is necessary to agree, in effect, that the marked disinterestedness of the Court during the course of May for the affairs in Germany had progressively accentuated this. The last letter from d'Argenson allows us to believe that the Council of Louis XV was quite indifferent. The news sent by Conti each day, growing steadily worse, his diplomatic and military considerations, his propositions to ameliorate an already bad situation, arrived at the camp below Tournai at the same moment when one was seeking the best means to pursue, in Flanders, an active campaign of certain conquests. What interest could then be served by a retreat on the Rhine, when the victor of Fontenoy was preparing a rapid and triumphant march on the Escaut? The role of the Army of Germany appeared ended while that of the Royal Army appeared that it would pursue one filled with brilliance and promise. Thus, the successive checks that Conti had suffered to embellish the brilliance of the success of Maurice de Saxe and the unfortunate result of the German campaign would augment the desire for happier results in the Lowlands.

It was, however, necessary to take some urgent precautions. On 14 July, as we have seen, General Bernklau had crossed the Rhine with a very important corps of Croats, Pandours, and other Freikorps. The populations of the Alsatian frontier were unaware of the direction taken by the Austrian contingent. Those of Lorraine and the Bishoprics were particularly concerned, being little assured by the bad reputation and the great mobility of Bernklau's soldiers.

De Creil had taken some precautions of their fears, but he had not succeeded.[282] Some spoke of 15,000 Austrians moving on the Moselle, others saw the Austrians approaching by Kaiserslautern or the Lower Saar, whose great dryness permitted an easy crossing to enter into Lorrain. Everything recommended the greatest vigilance to the governors, who sent the Lynden Hussars to the frontiers of the Bishoprics and the Bercheny Hussars to the Saar.[283] However as these troops, detached from the Royal Army, escorted at this time convoys of prisoners to Lille and that it was urgent to be diligent, it was ordered, on 26 July, to send two brigades of cavalry into the Bishoprics.[284] One

---

[281] Cf. Broglie, *loc. cit.* Vol. II, p. 99.

[282] Creil to d'Argenson, Thionville, 22 July (Corresp. A.A.).

[283] D'Argenson to Balincourt and to Criel, Borst camp, 24 July (Corresp. A.A.).

[284] D'Argenson to Maurice de Saxe, Ghent. 26 July. – The Minister added: "This detachment would place no obstacle to our plans for the rest of the campaign and there will be no detachment of infantry because this is what you have the greatest need of."

re-established the "lines" organized on the frontier of the Champagne and Thiérache. Once again, de Joyeuse received the command.[285]

A French spy had followed the march of the right wing of the Pragmatic Army during its difficult retreat over 12 and 13 July. He saw these troops defile on the morning of 14 July, at 9:00 a.m., at the moment when, crossing through the small village of Diligem, they marched on Vilvorde to cross the Brussels Canal. The strongest English battalions, he reported, did not exceed 450 men in number and the squadrons 100, 90 and even 80 horse. The Dutch had suffered no less; their battalions contained less than 450 men each and their squadrons ran from 150 to 160 horse, at most, with an average of 90.[286]

On 16 July, in a conference held with Königsegg, the Allies ordered the "disposition of the posts for the security of the Diegheim camp."[287]

Camping in two lines behind the Senne, from Vilvorde to within a cannon shot of the small village of Schaerbeeck, near Brussels, the Allies relegated their grand guards of cavalry on the rear, along the highway to Louvain. Each wing sent 800 men to Brussels and Antwerp. Relieved every eight days, these 1,600 men were to serve as a garrison in the two cities where there had been none. The right furnished the posts in Louvain, Vilvorde, and Pont Brûlé; the left at the Jacob, Trois Fontaines, and Trois Trous Redobuts, and the bridge at Laken. The pioneers and engineers put these posts into a state of defense. Free troops were placed at Grimberghe and Saint-Peters Woluwe; others left on an expedition towards Charleroi, as much to interfere with the requisitions of the French hussars and uhlans as to scout the left flank of the Allies and to pillage, when the occasion arose, all the part of the French frontier that the departure of Clermont-Gallerande for Chièvres had uncovered. Finally, they made the preparations necessary to flood all the ground between the Senne and

– In reality, one learned that Bernklau had recrossed the Rhine, on 24 July; however, these orders were not changed. One changed only the destination of the troops that one had sent to Maubeuge, Philippeville, Mézières , and Sarrelouis. Clermont-Gallerande constituted the two brigades and the army sent him two others [Crémilles to d'Argenson, Oordeghem camp, 29 July (Corresp. A.F.)].

[285] D'Argenson to Méliand, Beaupr, Creil, Joyeuse de Grandpré, Ghent, 27 July. – D'[Argenson to Méliand, Ghent, 4 August (Corresp. A.F). – One finds all the details of these true National Guards of the ancient régime in Colin. *loc. cit.* Vol. II, p. 173 and following.

[286] Unsigned letter from Brussels, dated 7 July (Corresp. A.F.).

[287] Waldeck, 16 July.

the canal from Evere to Vilvorde and Villebroeck.

The order of battle held in the Archives at la Haye gives the effectives of the Pragmatic Army in the field, at the end of July, at 53 battalions and 89 squadrons. These figures appear to indicate a slight reinforcement in the form of infantry since the beginning of the month. In light of the weakness of the units, this is only an increase in the number of battalions, not in the absolute strength of the army. "The enemy had no more than 40,000 men," reported the spy Mouvet, a bit after the same date.[288] The information of the French agent in Brussels exactly confirmed this information.

The strength of the garrisons in the fortress on the left flank of the Allies had varied little; but they still pursued their defensive activities with great activity. Ten battalions, some cavalry, and two free companies defended Namur with 100 cannon and 80 mortars. They still expected the reinforcement of material promised by the States. The defenders, however, counted no more than 6,000 men, as the Austrians were suffering from heavy desertion. Great distrust reigned in Namur. De Colliart, its governor, had sought to take control of the keys of the city, which had to that point been held by the bourgeois. These latter refused, adding that if the French appeared, they would open their gates so as to prevent the destruction of their city, "which had been rebuilt anew 35 years ago." Colliart had posted his Austrians on the ramparts; besides the Dutch refused them access to the citadel which they claimed to occupy alone. Despite everything, the governor tore down several houses in the northern and western suburbs to clear the edges of his fortress.[289]

With a view to preparing the resupply of Charleroi and Namur, the Queen had posted across the entire region an order directing the inventorying of all livestock. Upon the first request, a tenth was to be sent to Namur, at the expense of Their High Powers. Soon this same province was to see new workers being conscripted and 600 pioneers were to be furnished to the Pragmatic Army. Would eight Spanish sols and daily bread for the peasants torn from their land, at the moment of the harvest, suffice to compensate them for the time lost to this

---

[288]Mouvet to Maurice de Saxe, Philippeville, 20 July (Corresp. A.F.).

[289] Mouvet, Aulent, Thirion and Goderneaux to d'Argenson, from 14 July to 15 August (Corresp. A.F). – The first two were spies covering the fields, the third had an official function in the campaign of the Gates of Valenciennes. As to the last, he exercised, among other trades, that of captain of a free company of dragoons. All were excellent intelligence agents.

requisition?[290]

The garrison of Charleroi contained only three battalions, totaling about 1,500 men, and that only because of the desertion from the forces of Clermont-Gallerande during his stay at Maubeuge and Binch.[291] They worked actively on the city's fortifications, constructing new redoubts and mines. An Austrian engineer directed the work, but it was paid for by Their High Powers.

Mons was defended by 4,000 men, some cavalry, and some free troops, under the command of de Nava. An engineer was sent there from Luxembourg and "made a great noise upon his arrival, not having found the fortress to his liking," wrote Thirion. "He deployed a great activity, repairing the batteries, preparing the inundation, constructed the *chevaux de frise* of an extraordinary size, it was said, to be used in breaches. The habitual suppliers for the Austrians had received the order to organize magazines in the fortress. They only consented to obey on the condition that they were paid in advance. The magistrates of Mons worked to produce a remedy to this terrible situation; but, while waiting for it to be improved, the governor set about provisioning himself at the expense of the population on the neighboring French frontier who he periodically raided. The distance of Clermont-Gallerande provisionally favored these maneuvers, which would soon be brought to a halt."[292]

On the Allied right flank, the recent events as rapid as they were important had thrown great alarm in all the fortresses in this region. The circular letter, communicated by Maréchal de Saxe, on 14 July, to all the commanders of the Dutch fortresses, had not sufficed to reassure them. However, he "assured them of his desire do everything that would please them." He explained to them that he was obliged to send parties forward, but that he had ordered the detachment commanders to be certain that they did not penetrate into the lands of Their High Powers. To avoid any misunderstanding "he had the honor to beg the Dutch governors to place on the limits of their territory markers, emblazoned posts or any other marker that they thought appropriate."

Such were the Royal intentions, affirmed Maurice, and he was especially charged with watching over their strict execution. Despite this urbanity and this good will, the mistrust of the Dutch command-

---

[290] Mouvet to d'Argenson, Thyle-le-Château, 27 July. – Thirion to d'Argenson, Valenciennes, 12 and 13 August (A.F.).
[291] Thirion to d'Argenson, 14 July to 15 August. (*Ibid.*)
[292] Thirion to d'Argenson, Valenciennes, 13 August (Corresp. A.F.).

ers was not diminished and after having placed their border markers, they prepared their defensive depositions.[293] As the land of Waes had become the King's land, the situation of the Fortresses of Ecluse, Philippine, the gate to Ghent, Hulst, bordering this region, made the role of their governors very delicate. Flooding remaining the best of their defenses, the Dutch commanders had readied it. The populations, threatened with ruin by the invasion of water called on Louis XV, who greeted their complaints favorably.[294] To this source of conflicts were added the complaints raised by the rare incursions of the disinterested parties of French or Allies on the left of the Escaut. Du Chayla, Governor of Ghent, did not hesitate to threaten the Dutch governors and declared himself ready to find in these raids a pretext for a declaration of war against the Dutch.

Lieutenant Genreral J. Dibbetz, commander at Ecluse, requested instructions. Having no supplies, he brought some livestock into his fortress. Assailed by the complaints of the peasants, the requests of the small fortresses depending on Ecluse, and the incessant communications from la Haye, he did not know to whom he should listen. He said he did not have either the time to eat or to sleep. He had received some reinforcements, but he had only housed them with difficulty. Finally, the inhabitants of Ecluse had fled in great numbers.[295]

[293] "We should not allow ourselves to be surprised by the violent attacks that our neighbors have so commonly used (sic), if the desire one day obliges us to use the force of a very Christian arm. [de la Rocque to Waldeck, Hulst, 27 July (Corresp. G.H.)]. – Pierre de la Rocque, lieutenant general and colonel of a Dutch infantry regiment is also the governor of Hulst and the dependent fortresses, as well as the Commander-in-Chief of all Dutch Flanders, from the sea to the Escaut inclusive.

[294] Maurice de Saxe to de la Rocque, Alost camp, 8 and 11 August, Opdorp headquarters, (Corresp. G.H.).

[295] "It is incredible how much is lacking here; but it is not the good will that has caused this defect; but the power and the weather." [J. Dibbetz to Their High Powers, Ecluse, 26 July (Corresp. G.H.)]. – Other preoccupations no less grave disturbed this lieutenant general. After the capture of Ghent, Mölck in fleeing had hoped to seek refuge at Ecluse, thinking reasonably that Ostend did not offer him a sure defense. Dibbetz refused him entry into his fortress. The Austrian general complained to the Allies, who presented complaints to the Republic. The latter asked explanations of its general who hid behind his instructions, "because," he said, "the commissions of the governor stated that no troops should enter into the city nor leave without a license from the States, especially if they were foreign troops pursued by an enemy who sought any reason to break the peace, as du Chayla had already informed them." The Republic expressed its disapproval to its general, on 22 July, such that the latter found himself greatly embarrassed when the French troops, charged with ensuring the security of the King during his visit to Bruges, took up posts in the vicinity of Ecluse. Dibbets asked instructions in the case where the French pursued Allied couriers under

Henry Pichat

At Hulst, it was learned that the French were going to besiege Dendermonde: "As this place is only four long miles from here, the forts of the Escaut and us, when the winter comes, during the great freezes, we must hold ourselves on our guard," wrote the governor.[296] Th. Kemp, commander at Philippine, commanded only two companies of infantry and four master gunners. For some time already he had requested reinforcements. His superiors had contented themselves with doubling his garrison, but at this time, he demanded 4,000 men and 50 gunners. In waiting for the States to send him these troops, he very attentively observed the French movements and had extended the inundation in order to conform with a decision of the States dated 1 March 1745.[297]

On the Queen's lands, the alarm was greater still than in the Dutch fortresses. The officers of the Republic in garrison at Dendermonde complained bitterly to la Haye that the Austrian governor of this fortress had not yet completed the inundation, despite their pressing insistences.[298] There had been excellent reasons. The sluices had been very badly maintained. The fortress could barely count on the resistance of its works left in such a deplorable state. The old breach cut in the ramparts on the side of the Brussels Gate by Churchill's cannon, in 1705, were still open. The parapets were collapsing almost everywhere and in places one could knock down what remained with a stick. It being impossible to operate the sluices, Dendermonde lost the best of its defensive resources that came from its situation at the confluence of the Dender and the Escaut. The States had sent an engineer to restore the ruins, but he had barely been there eight days when he was ordered to go to Ostend: "Your Most High Powers," concluded a report dated 19 July, "can see, clear as day, that we are not in a position to defend ourselves from a sudden attack."

---

the walls of his post as they moved across the Waes region. "Must I allow then to enter, or no one? I see that without an order from Your High Powers, I must not permit access to the fortress of such a great number of enemies. I propose, therefore, that if they come too close, and in force, I shall ask them to retire outside of range of my cannon, if they do not wish to be fired on, at least as long as I have not received formal orders from Your High Powers to act otherwise." [J. Dibbetz to Their High Powers, Ecluse, 30 July (Corresp. G.H.)].

[296]De la Rocque to Their High Powers, Hulst, 4 August (Corresp. G.H.).

[297] Thomas. Kemp to Their High Powers, Philippine, 2 August (*Ibid.*).

[298] Lieutenant Colonel Van Casteel to Their High Powers, Dendermonde, 19 July (Corresp. G.H.).

Maurice De Saxe's 1745 Campaign in Belgium

At Ostend, the French had been expected to arrive at any time since 21 July.[299] If one believes Löwendahl, the English demonstrated a boastful optimism.[300] However, the Dutch addresses to the States presented a certain view of the reality in a less assuring perspective. These documents assure us that decay had rendered the circumvallation very difficult on the ramparts and by the sea winds filled the ditches with sand that the palisades crowning the parapets found themselves covered up. The Dutch cleared this away, but this work took much time. The city contained no casement, the bourgeois had fled, carrying with them everything they owned in barges, while movement by the canals remained open. The capture of Ghent had cut all communication by these routes. At Ostend too, the governor refused to complete the inundation. As for the garrison, it contained only 1,750 men and 6 inexperienced gunners at the end of July. This was the remains of a Dutch battalion, an English battalion of 400 men, a Scots battalion of 300 fusiliers, troops that had suffered greatly at Fontenoy and finally some fugitives gathered up by Mölck after Chayla were chased from Melle, Löwendahl from Ghent, and Dibbetz from Ecluse. Among these last troops figured some unusable Hanoverian cavalry, as the fortress contained no supply of forage. Supplies of food also were lacking. The governor could not hope for any assistance beyond what the London Court had requested. A colonel of the Guards came, gave a report and left, promising reinforcements that had not yet arrived at the end of July.

---

[299]Lieutenant Colonel C. de Saint-Amand to Their High Powers, Ostend, 21 July (*Ibid.*).

[300]"The French wish everyone to believe by their preparations and by their movements that they are seriously resolved to go before Ostend and to undertake a siege of this important fortress. They have advance the largest part of their forces on the side of Ostend. There is already, beyond Bruges, a corps of 30,000 men, all their other troops moving along the same road. Who can as a result contest their design on Ostend, considering above all that they have with them 80 heavy cannon and that, according to a letter from Paris, 12 ships of the line and, according to other French letters 20 to 30 have the order to second the operations of the army destined to make the siege of this fortress? However the men who know the situation of the fortress defended on one side by the sea and on the other by flooding, are informed that it is abundantly provided with troops and all that which they might need and who know how important this fortress is for the English and Holland, whose naval and land forces are now in a good state to prevent it from falling into French hands, still doubt the truth of this rumor or they are less persuaded of it, in case that this enterprise occur against all the expectations and know there is a great gulf between undertaking a siege and the capture of this fortress." (*Memoirs of Löwendahl, loc. cit.* p. 84.).

As for von Waldeck, in the middle of all this disorder, he continued his diplomacy. The equality of the garrisons was no longer in question. This had provided the subject of many irritating and sterile discussions in June and at the beginning of July. Now other concerns preoccupied von Waldeck. He waited to see if the French would only besiege Dendermonde and watched their immobility for nearly 20 days, in the Borst camp, which appeared characteristic to him. "This confirms to me," he said, "that the enemy [French] would not make much progress during this campaign."[301] The Powers no doubt did not share the illusion of security of their general, because, on 28 July, they enjoined him to take all the measures necessary to guard the west bank of the Eschaut, that the anticipated fall of Dendermonde would not fail to uncover.

As a result, Waldeck had sent to Hulst all the useful recommendations; but de la Rocque responded by informing the Prince of the bad state of the posts covering the left of the Escaut. In particular, the forts at Marie and Perle demanded urgent repairs. As these works found themselves occupied by the Austrians and that according to the Treaty of the Barrier of 1708 and the la Haye Convention of 1717, they were to pass into the hands of the troops of the Republic, Waldeck imagined, in his zeal for the interest of the States, to repair the forts in a bad state at the expense of the Queen. He announced this project to Königsegg who counseled him to speak to von Kaunitz.[302] This diplomat showed himself much less easy to convince than the old Austrian marshal and the Dutch to resign themselves to occupying the forts of the Escaut in the dilapidated condition in which the Austrian engineers had left them.

In sum, the greatest alarm that reigned everywhere was further increased by the general disarray. It was necessary to convince them otherwise that if the Republic showed itself little assured, the proximity of the victorious French army increased the fears that the attitude of the Allies was not in a nature to calm. Would the Dutch declare war when the French camped near their frontiers? They still had 55,000 line infantry, this Royal army, on 1 August and only detachments were employed to guard the new conquests that had little

---

[301] Waldeck to Their High Powers, Schaerbeek Camp, 2 August (Corresp. G.H.).

[302] Wenceslas Antoine, Count von Kaunitz Rittberg, of the Holy Roman Empire, Lord of Essens, Stedsdorf, and Wittmund, etc., Intimate Councilor of State to Her Majesty the Queen of Hungary and Bohemia, her Minister Plenipotentiary for the Government of Her Lowlands. (Corresp. G.H.).

diminished since 1 July.[303]

No one foresaw the direction in which Marshal de Saxe would send his troops. "I do not believe that the siege of Audenarde will hold us up long," he said, on 18 July, "afterwards, we will see."[304] He was master of the Upper Escaut and Ghent; the Allies, weaker than ever, remained indecisive in the defense behind the Brussels Canal and the Zenne River, while only two months of good weather remained to roll out. The moment to undertake new operations appeared to have arrived. What would one do?

"The goal of a war is ordinarily more political that military, but the goal of each campaign and each operation must be to procure advantages and to do damage to one's enemy," professed Maurice de Saxe in his report of December 1744. What would be the practical manifestation of this concept? Definitively, there could be no more question of a return to the offensive on Mons, Namur, or Charleroi. The resolution to move the war on the Escaut prevented such a project. Two other interesting solutions presented themselves.

At this time, the Dutch provided more than half of the effective forces of the Pragmatic Army then in the field in Flanders, without considering those that it also maintained in Germany.[305] Maria Theresa didn't appear disposed to diminish the charges, each day more heavy, supported by the States General. Despite all the sacrifices endured for the common cause, the Republic could fear a declaration of war followed by the immediate invasion of its territory. Was the election of Francis of Lorraine to the Imperial dignity sufficient compensation? Could the French not hope to detach Holland from the Coalition if, after having captured Dendermonde, the French Army came even closer to the Republic's frontiers?

In supposing that the execution of a parallel project would be crowned with the most complete success, it presented, however, certain difficulties. Maurice could not move in the direction of Antwerp without uncovering Audenarde, Ghent, and Dendermonde in light of assuring the security of his line of communications. It was import-

---

[303] The 7,756 infantry in hospital, on 1 July, were no more than 6,717 on 1 August [Situation of the infantry on 1 July and 1 August (Corresp. A.F.)].

[304] De Saxe to Folard, Borst Camp, 18 July (Corresp. A.F.).

[305] The States General maintained, in Germany, 14 squadrons and 6 battalions, totaling 8,000 men. It is appropriate to add to that the 6 reserve battalions. Though they had not yet left Holland, these latter troops were nonetheless stood up. [State of the Dutch troops that were in the Allied Army in the 1745 Campaign on the Lower Rhine. (Arch. L.H)].

ant, also, that as he moved, he leave the necessary forces to protect Maritime Flanders. Ostend and Nieuport would become useless for the English to communicate with the Continent; however, the possession of these two fortresses still gave them the means to attempt a diversion on the French northern frontier. Finally, Maréchal de Saxe could not march on Antwerp without preventing the Pragmatic Army from threatening his right flank. He had to repeat against it the threats he exercised the previous month. What then would be the attitude of the Allies? If the Brussels Canal and the Zenne River offered only an illusionary defense against the 50,000 French soldiers, the value of these natural obstacles would be augmented by the diminution of the French numerical superiority. In addition, the Rupel would subsequently furnish the Allies with a new rampart in the case where, in refusing combat, they withdrew in order to draw the French as far away as possible from the Upper Escaut. One did not have to hope to capture Antwerp by renewing the maneuver that was so successful against Ghent. In sum, an action engaged on the first of these places could engage the French in a series of operations that would, no doubt, not be very difficult. However, it risked lasting a long time and did not offer, in any case, more than the capture of Antwerp, which could not be held as winter quarters.

Another solution, to the contrary, permitted one to envision what could be a rapid result, but a result that would be certain and important coming from a sure blow. From the opening of the 1745 campaign, Flanders had become the principal theater of the war. The part engaged in Germany appeared lost. One of the happiest consequences of the victory of Fontenoy was to have moved the French efforts to a battlefield where, at least, the French fought for themselves. In reality, the naval powers had thought to find in this war excellent operations to resume the fight with France that had been interrupted in 1723. The campaign of 1744, as well as rapid and successful operations of May and July singularly reduced the hope of the Anglo-Dutch. The time of the complete ruin of their illusions appeared to have come as well as that to repair all the damage done to the French northern frontier by the last series of treaties. Since the powerful and victorious French Army held the heart of the land, only having before it a weakened and divided army, the sane strategic doctrine counseled the French to take full advantage of such a favorable situation. The capture of Ostend and Nieuport, which would assure the security of French Maritime Flanders, that of Dendermonde, which gave them the course of the Dender as a shield for their winter quarters, could only be the normal conclusion of the campaign on the Escaut. Two months of good weather remained in which the troops could pursue their operations with every facility. The actual position of the army, particularly favorable, permitted the covering of the projected sieges while holding superior forces united before which the Allies could only remain impotent.

On 25 July, the captured garrison of Audenarde left for Lille, passing before Louis XV on the fortress' glacis. His Most Catholic Majesty visited, at this time, his new conquests. His trip proceeded under the protection of a large number of troops drawn from the fortress garrisons, the Melle camp, where the Prince de Pons commanded, as well as the army. The King's Household, the Gendarmerie, and the Guards, that is 21 squadrons and 9 battalions, had left Borst to accompany the King.[306] The Prince de Pons could not take more than 34 squadrons and 4 battalions from the French lines without leaving the Quatrecht bridge, that is to say, the direct communications of the army with the left of the Escaut and the Waes territory, at the mercy of an Allied expedition coming suddenly out of Dendermonde or Antwerp.[307] In these conditions, how could one organize the operations decided on towards Ostend, Nieuport, and Dendermonde?

Considering the importance of these three fortresses and the reasons for capturing them, the sieges of the first two were to be pursued as soon as possible. Dendermonde could fall last, without any inconvenience. It was, nonetheless, preferable to profit from its bad condition to attack it quickly, giving the Allies neither the time to repair its ramparts, nor to reinforce the weak garrison that defended it. While reducing, as it was appropriate, the illusory optimism of the English news, which misled no one and even less Maréchal de Saxe as he had been well informed on Ostend and Nieuport for a long time by d'Aunay, the commander at Dunkirk; without dividing the adverse opinions which claimed these fortresses would be easy to reduce, the sieges of these two cities appeared that they would be difficult. Some wrote, "that there was not a merchant in Ghent, Bruges, Tournai, Lille or Paris that was in not able to report that the fortresses in Ostend and Nieuport were greatly neglected." Other individuals, well informed, considered that these sieges would be "anything but normal." The In-

---

[306] Dispositions for the security of the march of the King to Ghent, to Bruges (Corresp. A.F.).
[307] This force did not return to the French lines until 4 August (Pons, 4 August).

Henry Pichat

tendant Sèchelles, who surely knew everything regarding the precautions taken, said to the Count de Clermont: "The enterprise is begun; I reserve my judgment on its success."[308] The geography of Ostend and Nieuport furnished most effective means of augmenting the resources of a well resolved governor. The sea and the canals permitted the movement by water of a considerable force. One no longer doubted that the London Court was not resolved to co-operate as much as possible in their resistance, be it by sending reinforcements, be it by furnish the support of a relief fleet.

Were the French going to simultaneously constitute two distinct siege corps for the operations on Ostend then Nieuport on one hand and on Dendermonde of another? Such arrangements would reduce the effectives of the army to at least 30,000 men. Because of the anticipated difficulties, few intrigued to dispute, with Löwendahl, chosen by the King, the favor of directing the attacks on Ostend and Nieuport. However, many sought to see themselves designated to direct the siege of Dendermonde, an operation that was more certain and would be pursued under the King's eyes. It appeared, in effect, that the Maréchal de Saxe would oppose the formation of two distinct detachments, which would hurt the numerical superiority of the French Army over that of the Allies.[309]

Irrespective, Löwendahl received the command of a detachment of 14,000 men which were to take Ostend first, then Nieuport. At the same time, Lieutenant General d'Harcourt was designated to besiege Dendermonde with the troops necessary furnished "day by day and according to the needs of the army." The army was to camp on the right bank of the Dender near Dendermonde in such a manner so as to cover the operations undertaken by Löwendahl against Ostend

---

[308] R.A.- R.A'. – Pons, 4 August. – Sèchelles to Count de Clermont, Alost, 5 August (Corresp. A.F.).

[309] R.A. said to this end: "The King had decided to leave Ghent. However, he changed his mind, and no one knew why. The detachments for Ostend and for Dendermonde were formed in his presence, to the great regret of the general." The preparations (described by the *Journal of the Lawyer Barbier* and the *Memoirs of the Duke de Luynes*), made in Paris to receive the King, confirmed the first part of R.A.'s assertion. The Prince de Pons furnished a very characteristic indication when writing, in his journal, on 31 July: "The King will arrive today at Ghent and at Bruges. His return decided our movements. There was a project formed, but no disposition that clarified it. There is a strong rumor that Prince Edward has landed in Scotland, this news must have thrown the English into great perplexity. We await news on the impact that this landing will produce in England."

and Nieuport.[310]

On 27 July, de Saxe prepared to march forward to move the army on the Dender. During the evening, the six crossing points over the Swalme prepared for the march of the next day were given a post of 10 fusiliers and a sergeant, who were charged with preventing the crossing of any equipment other than that of the Minister, from crossing before the troops.[311] The food caissons collected to gain, as was their habit, the best roads. At 1:00 p.m., the 58 squadrons and 83 battalions of the army were camped at Oordeghem.[312] The troops of the Prince de Pons, in their camp at Melle, had covered the left of this march.[313]

On 2 August, de Saxe prepared to end his movement. The 12 companies of grenadiers from the Piedmont, Bouzols, and Beauvoisis Regiments left for Alost, led by an infantry brigadier, and took up a post in this city.[314] This officer "was advised that the troops must observe the most strict discipline and that the King's sergeants, with the quartermaster corporals of the army, would join them that evening to visit and mark in Alost the lodgings for the King's quarters, those of the Princes, and the headquarters."[315] One awaited the return of Louis XV to the army.

On 3 August, at 4:00 a.m., the encampments were formed as on 28 July. The army marched in six columns, the first and third having a general rearguard formed with the old guards relieved at 5:30 a.m. The wings contained 32 squadrons, 8 battalions on the right and 26 squadrons, 4 battalions on the left. Columns 2, 3, 4, and 5, and that of the center were formed entirely of infantry divided into 20, 17, and 20 battalions and a convoy. As previously the troops of each column brought with them their baggage escorted and followed by a rearguard.

Early on the morning of the 3rd, the army found itself camped after a march that was executed in good order.[316] It had arrived with only 69 battalions and 58 squadrons. The 14 battalions from Eu, Bet-

---

[310]Brézé. – Pons, 8 August.

[311] March order of the 28th (Coresp. A.F.).

[312] Two brigades of cavalry remained in the camp at Borst. They had been sent to Chièvres, as we noted at the beginning of this chapter.

[313] All the details of this march are furnished by the above cited march order of the 28th.

[314]Movement order for the march of 3 August (Coresp. A.F.).

[315] March order for 3 August (Coresp. A.F.).

[316] Brézé, 4 August.

tens, and Séédorf designated for the siege of Antwerp remained until they received new orders at the Oordeghem camp. Camped in two lines from Hofstaedt to Nivone, on the right bank of the Dender, the troops reserved in their lines the intervals necessary for the Guards and the contingents that the Prince of Pons would bring forward from Melle on 4 August. The Piedmont, Bouzols, and Beauvoisis Regiments camped on the right bank of the Dender before Alost where the King took up his quarters during his return on the 4th. The hussars occupied Ninove and an equally important detachment was established in the Afflighem Abby, in order to watch the highway from Brussels.[317]

On 30 July, upon learning of the movement of the French Army, the Allies no longer doubted the pending siege of Dendermonde. They immediately thought to reinforce the garrison of this fortress which was then defended only by two battalions. Desertion had reduced the garrison to about 700 men. The Allies had not decided, however, on a reinforcement until the 3rd and by what means! They sent, according to von Waldeck, 600 infantry (300 Dutch and 300 English) and 10 gunners. This detachment was to embark on boats at Antwerp and move up the Escaut to Dendermonde.[318] No doubt suffering from the desire to escape all responsibility for such a late decision, the Prince added: "I was not consulted at all on the manner by which these 600 men would be sent to Dendermonde and no precaution was taken for their security."[319] The French did not have, however, at this moment on the left bank of the Escaut more than a few parties of detached Grassins moving downstream from Dendermonde with Colonel de Vault and 1,000 infantry.[320]

Assembled at Antwerp during the evening of 4 August, the Allied reinforcements embarked. They were commanded by Major Harel, "a man of merit and a very good officer," according to the judgment of von Waldeck. Harel and the Dutch left about 700 p.m., in five boats.[321] The English were to follow after an hour. Before departing, Harel prescribed sailing to secure and reach Dendermonde

---

[317] All the details of the march and the installation are provided by the movement order.

[318] Waldeck, 3 August.

[319] Waldeck, 5 August.

[320] "We erected a boat bridge at Baesrode where de Vault, with 1,000 infantry, crossed and took up a position at the confluence of the Dendre and the Escaut." Brézé, 4 August.

[321] Report to the States on the battle of Saint-Amants. – Transmitted by Waldeck (A.L.H.).

no matter the cost.[322] At 11:00 pm., the group of Dutch arrived at Fort Sainte-Marguerite. It dropped anchor waiting for the tide before resuming its movement. It reached, without obstacles, a point about a half mile downstream from Saint-Amants. At this time, local peasants cried to Harel that 1,500 French soldiers occupied the village. The major had his soldiers lay down in the bottom of the boats and continued on his route.[323] This subterfuge did not fool the vigilance of the Grassins' advance post watching the entrance to the village. Upon their arrival, the embarked Dutch were greeted with a heavy fusillade. They passed without responding.

A bit upstream from Saint-Amants, the Escaut presented a bend that was difficult to pass and at that point, on the right bank, there was a dike behind which the Grassins had placed 300 of their arquebusiers. They had also advised the pickets of the detachment of de Vault in order to block, at any price, the road to Dendermonde to these suspicious embarkations which the first musketry had not stopped. At the moment when the Dutch arrived before the dike, after having crossed the village, a new discharge greeted them. A boat, badly steered by its panic-stricken helmsmen ran aground in the reeds on the right bank so close to the dike "that one could touch it with a spontoon."[324] The incoming tide pushed the two other boats from the shore at the same moment when the Grassins' musketry killed one of the pilots. Drifting in their turn, these boats became stuck in the mud alongside the first. As for the last two, bearing Harel's body, who had been killed, they crossed the dangerous passage under a hail of balls and sailed towards Dendermonde.

For an hour the grounded Dutch attempted, in vain, to refloat their boats under the Grassins' fire. They were working their hardest when they saw the two boats that had succeeded in passing returning down the river. The tide and discouragement drove them back and, moving along the left bank, they regained Antwerp.[325] Besides, the alarm had been raised. Certianly they could not have passed Baesrode. Their crews had suffered cruelly from the fire of the Grassins. They escaped, all the same, because the attention of the Grassins was directed on the shipwrecks. For another three hours, without becom-

---

[322]*Ibid.*

[323]According to his report, Harel had to employ violence to compel his pilots forward, "sword at their throats, in order to not separate one from the other."

[324] Report to the States, *loc. cit.*

[325] *Ibid.*

ing discouraged, the grounded Dutch attempted to rejoin their fleeing comrades. Finally abandoned, and not seeing the English boats
coming that might have helped them, threatened with being burned
in their boats, they surrendered. No resistance was possible because
about 30 tartanes [type of boat] loaded with French soldiers arrived
from Baserode. In consideration of their brave conduct, de Grassin
gave the Dutchmen the honors of war. The capitulation specified that
they would be exchanged on the first occasion.[326] Maréchal de Saxe
ratified these conditions. By virtue of the obscurity of the area in the
middle of which the action occurred, the Dutch only lost 11 dead.
Otherwise, the fire of the French soldiers, who one "did not see any
more than their heads extending over the dike", had obliged the Dutch
to remain in the bottom of their boats.[327] The latter had vainly attempted to place some men on the upper deck to repost. However,
they had to renounce this effort, because the Grassins "fired accurately
at anyone who showed themselves."[328] The French lost 20 dead and
wounded and they took 180 prisoners. The five English boats had not
yet arrived, even though they had left at the agreed upon hour. The
result of this adventure was that, on the 6th, the Duke de Chevreuse left
the French camp with four regiments of dragoons.[329] He established a
bridge over the Escaut downstream from Dendermonde and camped
that same day at Grimberghen.

After its installation in the Alost camp, the army had completed the construction of three bridges over the Dender, the first, between
Erembodeghem and Alost; the second, between that city and Hofs-
taede; the third, constructed only of fascines, and a bit downstream
from the latter location. They were to serve on the 7th, for the movement of the army on Dendermonde. As the attack on this fortress was
to be directed from the west, between the Dender and the Escaut, de
Saxe had decided to camp on the right bank of the river, from Appels
to Ninove.

To occupy this new position, the army marched out at 6:00
a.m., on the 7th.[330] The Carabiniers, with their baggage, the camp parties, and the new guards departed first, at 5:00 a.m. Dendermonde
found itself invested by 10 squadrons of carabiniers established be-

---

[326] Capitulation annexed to the report.
[327] Report to the States, *loc. cit.*
[328] *Ibid.*
[329] Bréze, 11 August.
[330] March order of 7 August (Corresp. A.F.).

tween Appels and the fortress outside of cannon range. The army formed in five columns following the troops that formed the advance guard. After having crossed the Dender, over the recently established bridges, it occupied a camp marked out in two lines from Wiese to Baesrode, moving by Lebbeke. The Army Headquarters and the Royal Headquarters, the treasury, and the mobile hospital remained in Alost. The infantry brigades that had previously covered this city when leaving were replaced by the Gendarmerie and the Guards.[331] This short march, or better still, a simple displacement, was completed peacefully. Besides, from 7 July, the Prévôté de Hôtel destined to escort the King's baggage had been reinforced. The many disorders that occurred during the first marches of July had emphasized the insufficiency of the forces charged with containing the convoys [of baggage]; one augmented it with a brigade of constabulary. During the duration of the march, the Grassins scouted the ground between Saint-Amands and the Affighem Abbey, searching with care the Buggenhout Wood and Afflighem. Detachments occupying Ninove and the abbey remained where they were. Finally, to assure the security of such an extended front, the Grassins received the order to continue their exploration that had begun on the 7th. As for Lieutenant General d'Estrées detached to the right with 1,500 horse, he was to, for the same reason, patrol to the north and south of the Alost to Brussels highway and to link his reconnaissances with those of the Grassins. [332]

After having installed his headquarters at Lebbeke, Harcourt began his operations against Dendermonde, on 8 August. The flooding, above all, constituted a serious defense on the side chosen for the attack. At its narrowest point, the flooding covered a distance of about 200 toises [400 meters] before the walls. The roads to Alost, Brussels, and Malines, very narrow, alone crossed it. Several masonry redoubts cut the two latter and commanded the sluices. "This inundation," recounted Brézé, "was formed by the holding back of the waters of the Dender held by a great sluice that at low tide held back 6 pieds [2 meters] and at high tide held 13 pieds [4.3 meters], and was augmented by several small sluices that daily served to hold back the waters of the smaller tides."[333] If a parallel system had been conserved in a perfect state of functionality, the inundation could only have been

---

[331]Special order for the Guards, 7 August (Corresp. A.F.).

[332] Bréze, 11 August.

[333] *Ibid.*

drained progressively. The French hoped, to the contrary, to drain it quickly and very successfully, because then the fortress would not fail to quickly fall.

During the night of 8/9 August, the engineers constructed a battery with four guns on the edge of this sheet of water in order to break down the redoubt closest to the Malines highway. Without waiting for the effect of the artillery, the French grenadiers seized this work, "sword in hand" before daybreak.[334] They held out there, despite the violent fire of the garrison and broke open the dike on the right. The waters of the flood flowed into the Escaut. On 10 August, the highway to Malines and its approaches were entirely drained.[335] To the right of the road to Alost the engineers established six cannon and four mortars and meanwhile one could advance along the Escaut, downriver from Dendermonde as far as a sluice, whose sluice gates were opened.[336] It was soon necessary to close them because the tides "brought as much water as they drained out with they were open" and the sluice found itself too exposed to the garrison's fire for one to maneuver the gates at will during the day. With an eye towards rectifying these disadvantages, new openings were opened in the dikes along the Malines highway. On the 11th, the French artillery opened an enfilade fire along the defensive front and during the following night, a large part of the flooded ground found to be dry, 2,700 workers opened a trench.

On the right wing of this parallel, which was completed, a battery was constructed to destroy the second redoubt on the Malines Highway, but the garrison abandoned this work. In the center, the French prepared two other batteries with an eye towards bombarding the advanced works on the Brussels Gate. On the left a flying sap was pushed forward with an eye towards turning a lunette that barred the access road to this gate. At 5:00 a.m., on the 12th, all the artillery of these works opened a heavy fire preparing for the attack by the French infantry.[337] "The besieged," wrote Brézé "saved us the effort" and raised the white flag. After the habitual discussions renewed with each capitulation, the Austrian Governor Tenderfeldt surrendered and on the 15th the garrison departed. Only 600 men guarded Dendermonde,

---

[334] *Ibid.*
[335] D'Harcourt to Maurice de Saxe, the camp before Dendremonde, 11 August (Corresp. A.F.).
[336] *Ibid.*
[337] D'Harcourt to Maurice de Saxe, Lebbeke, 12 August (Corresp. A.F.).

which the Marshal intended to dismantle. This justified the epitaph
coming from the pen of a contemporary: "There remained with the
garrison sufficient material to delay us for six days; it had not held out
a minute after the surface of the ground was cleared of the water."[338]

While d'Harcourt so easily captured Dendermonde,
Löwendahl pursued before Ostend the most important and most difficult siege of the campaign. Eighty battalions and five squadrons had
been designated to form the detachment given to Löwendahl, totaling
about 14,000 infantry and 700 horse. On 4 August, Brézé sent from
Tournai the artillery and ammunition necessary on boats directed towards Bruges. One had, in effect, chosen this latter city as the supply
center for the siege corps, because of its position at the intersection of
many canals and also because this region contained sufficient wood
to construct fascines and the freshwater indispensable for the troops.

The approach marches of Löwendahl's troops were rapid and
organized in such a manner so as to expedite the investment of Ostend.
Nine battalions were united at Ghent.[339] They departed on 4 August,
sleeping at Aeltere and reaching Bruges where de Contades had taken
command.[340] They left this city on 6 August and reached Oudenburgh
the same day at 5:00 p.m. There some peasants informed them that
an engineer and some soldiers had left Ostend in a boat in order to
break the dikes of the canal between that city and Oudenburgh. Contades sent two companies of grenadiers, who killed the engineer as
well as 24 men, and chased away the rest of the Allies before this
terrible project could be executed. The following day, Contades left,
leaving the battalion of Saint-Brieuc at Oudenburgh to furnish workers to make fascines as well as to guard the canal and the Zantwoor-
de bridge during the course of the siege. During the course of their
march, two dragoon squadrons joined them.[341] They left Liffinghe to
block the land communications between Nieuport and Ostend and finally reached Mariekercke.

The 14 battalions remaining in the Oordeghem camp with
Löwendahl departed on 4 August.[342] Sleeping, on the 5th, at Aeltere,

[338]R.A'.

[339] These battalions consisted of the Grenadiers Royaux, Saint-Breuc Militia, Crillon
and Laval Regiments, forming, respectively, 4,1, 3, and 1 battalions.

[340] He commanded at Bruges from the occupation of this city and was replaced by de
Courbuisson.

[341] These squadrons came from the Beauffremont Dragoons, coming from the Melle
camp by Bruges.

[342] Formed by the Eu, Löwendahl, Bettens, Wittember, Seedorf, and La Cour-au-

and the 6th, at Bruges, these troops arrived at Oudenburgh. Löwendahl found his headquarters installed there by de Contades. Löwendahl counted on the arrival of his artillery in a few days. He detached Bettens from his column and sent him to Maréchal de camp La Marck, who had left Bruges, on the 7th, with the three remaining squadrons of the Beauffremont Dragoons and they reached the dunes at Lissemoris the same day. Once the artillery arrived, La Marck was to receive six cannon and eight mortars. He was to arm the batteries that he was to construct at Lissemoris in order to bombard the port of Ostend and its entrance.

On the side of Mariekerck, the rapid march of Cotnades had surprised the Allies as they organized their defense. The Governor of Ostend had only just begun putting the small Fort Albertus in a state where it could offer some resistance. It was too late to undertake such preparations. The French occupied Albertus without a fight.[343] This work became the point of support for the left of the French camp traced out in two lines from the sea to the Nieuport Canal. On the 8th, the French troops had completely installed themselves. A mobile hospital was established at Liffinghe, thus maintaining the ability to easily evacuate the transportable sick and wounded to Bruges by the canal.

Löwendahl had, in sum, virtually achieved the investment by the 7th. On that day, at 4:00 p.m., inspecting his communications by water, escorted by four battalions, he captured Fort Plasschendael, which was constructed at the junction of the Ghent and Nieuport Canals. Solid and well-armed, this work "was not a place to be taken by assault, but there was only a soft defense by the Dutch, which allowed this to happen."[344] Two battalions marched by the right bank of the Plasschendael Canal, which continued to advance on the fort which Löwendahl summoned as soon as the French troops were within musket shot. The Dutch officer immediately surrendered with 72 men and 10 cannon. The French lost 3 men.

Contrary to expectations, the artillery, which had left Tournai on the 4th, had yet to arrive. Surprised by this delay, Löwendahl sent a messenger to Bruges to press the forwarding of the necessary material.[345] In awaiting the arrival of his artillery, he began the attack works in order not to lose any time. He changed his preliminary dispositions.

Chantre Regiments, counting respectively 2,2,2,2,3, and 3 battalions.
[343] Löwendahl to Maurice de Saxe, Oudenburgh, 7 August (Corresp. A.F.).
[344] Löwendahl to Maurice de Saxe, Fort Albertus camp, 8 August (Corresp. A.F.).
[345] *Ibid.*

Maurice De Saxe's 1745 Campaign in Belgium

He moved his headquarters to Mariekircke and directed that upon the arrival of the artillery part be landed at the Snaeskercke bridge and part at Liffinghe. With horses the French transported to Lissemoris, to the north of Ostend, the material assigned to de la Marck's batteries and to Albertus that which Löwendahl had personally reserved for it.[346]

Events delayed, until the 13th, the execution of this order, and by the first effect of this unfortunate delay, the English were able to send reinforcements into Ostend. However, from where did this reinforcement come? D'Aunay, commander at Dunkirk, had indicated its approach without indicating its source. Certain preparations made in England had, it appears, succeeded in the formation of a squadron, which, on 2 August, put into port at Margat.[347] From where were these 14 ships detached? Did they come from Antwerp? If not, it is otherwise, very difficult to establish the importance of this resupply. Some have evaluated the number of soldiers thrown into Ostend at 1,400; others at only 650.[348] Be it what it may, deprived of his cannon, La Marcke could do nothing to block this movement.

The nature of the terrain, the system of the fortress' defenses, only permitted it to be turned by the southwest front. The need to cut communication over land with Nieuport had also determined the emplacement of the French camp. In these conditions, Löwendahl would conduct his attacks in the region of the dunes, taking the road from Ostend to Nieuport as the axis of his movement.

Fixing in the ordinary point of departure of future movements about 160 toises from the fortress, Löwendahl began to chase back inside the wall all the garrison parties that had remained outside of it.[349] The garrison had opened fire from all its redoubts and the ships in the roadstead took part in a thunderous, but ineffective concert.[350]

---

[346] *Ibid.*

[347] Maréchal de Belle-Isle to d'Aunay and d'Argenson, June, July, and August (Corresp. A.F.).

[348] "According to the word that I received from the city, this convoy was charged with food and war supplies, artillery, and from 1,300 to 1,400 men." (Löwendahl to Maurice de Saxe, Fort Albertus camp, 8 August (Corresp. A.F.)],

[349] Löwendahl to Maurice de Saxe, camp before Ostend, 9 August (Corresp. A.F.).

[350] "The enemy fired on us today over 400 shots without killing or wounding anyone." [Löwendahl to Maurice de Saxe, camp before Ostend, 9 August (Corresp. A.F.)]. As to the English ships, on the 13th, there were only two and their artillery was quiet. The rest returned to Antwerp to put into security everything that the English had evacuated from the Ostend magazines.

Henry Pichat

The first result was the provision of 40-50 cannon which the besieged had placed in line on the attacked front, so Löwendahl asked for a reinforcement of material, which was granted to him.

At 4:00 p.m., on the 10th, the workers in the trenches began to remove the dirt and at the same time they encountered their first and a serious difficulty. As they dug, they encountered water at a depth of two feet.[351] This required that the parapets as well as the revetment of the batteries be made thicker to remedy the insufficient protection resulting from the impossibility of digging deeper. This problem had not been foreseen. The depot of gabions and fencing organized at Bruges was soon exhausted. As a result, one had to send out again the workers installed there where there were woods; that is to say, along the banks of the canal, not far from Bruges, over a distance of three miles from the camp. The nature of the ground, which was very soft, did not permit moving these fascines by wagon. It was necessary to carry them on the backs of men over a distance of a half mile. The pioneers and the Saint-Brieuc Militia were not sufficient, and the siege corps had to furnish many workers. This resulted in an appreciable loss of time in advancing the works as well as great fatigues for the troops. If the first problem was a minor problem, the cannon had still not arrived, so the beginning of the siege was delayed. Löwendahl prudently requested that he be sent a brigade of infantry.[352]

During the night of 11/12 August, the garrison broke the dikes on the Nieuport Canal. This uncorked the water that they had held back, but as the governor of this fortress closed his sluices, the flooding threatened to cover the French trenches. The French then cut two holes in the left of the Ghent Canal to draw off the waters.[353] The execution of these works and the violent rain that had begun to fall did not stop the work, except on the 13th, on the two parallels that were finished, even though the garrison maintained a heavy fire.

Finally, on the 13th and 14th, the long-awaited artillery began to arrive. Expedited from Tournai to Bruges, on 4 August, it did not leave Bruges until the 13th. At the moment it began moving on Oudenburgh, it was discovered that the boats on which it had been loaded were too

---

[351] La Tour (Chief -of-Staff to the siege corps) to d'Argenson, camp before Ostend, 11 and 12 August (*Ibid*).

[352] La Tour to d'Argenson, camp before Ostend, 12 and 13 August (Corresp. A.F.). – The Beauvoisis Brigade (4 battalions) left the Alost camp on the 16th and arrived on the 19th at Fort Plasschendael (Army Situation of 13 August. *loc. cit.*).

[353] Löwendahl to Maurice de Saxe, camp before Ostend, 12 and 13 August (Corresp. A.F.).

large to fit through the canal locks. It then had to be transshipped onto smaller ships, but, as one had only a limited number of these boats, a shuttle was organized. Despite the execution of all possible efforts, the last convoys did not leave Bruges until 15 August. During the morning of the 15th, peasants informed the French that the dikes on the Bruges Canal were to be cut.[354] The rains and the action of the high tides had caused a disastrous accident which had drained the canal over which this precious material was being moved.[355] At the price of tremendous effort, the French moved to repair the damage. The delay in the arrival of the material, the growing fatigue of the troops, the persistent bad weather, and finally, a last alert determined Löwendahl to press the siege that so many unfortunate circumstances conspired to extend.[356] Besides, this gave the artillerists all the time to make their preparations. During the night of 16/17 August the French began arming the batteries. This was the cruelest period of the siege.[357]

During the preceding days, the garrison had consumed so much ammunition that its fire now relented, when, during the morning of the 16th, a number of ships entered the roadstead with the rising tide. They brought with them a new supply of ammunition, which was completely landed by 4:00 p.m. Immediately the garrison's fire resumed all its violence such that the French batteries suffered greatly. By virtue of the rain, which softened the ground, this operation immediately became more difficult. It was necessary to move the guns by hand. The night was not sufficiently long as the guns became bogged down. They were momentarily abandoned during the day after they were covered with fascines. Finally, on the 18th, at the cost of much fatiguing labor by the troops, Löwendahl opened fire with 28 cannon and 80 mortars. The following day La Marck's artillery joined the action. Their first projectiles sank the last English ship as it moved out of the port. The other ships had been able to withdraw as peacefully as they had arrived.[358] Resolved to finish during the night of the 22nd/23rd, Löwendahl decided to make "a greater effort" because he was able to link a flying sap to the salient angle of the covered road of the fortress

[354] Courbuisson to d'Argenson, Bruges, 13, 14, and 15 August (*Ibid.*).
[355] La Tour to d'Argenson and Löwendahl to Maurice de Saxe, camp before Ostend, 15 August (Corresp. A.F.).
[356] *Ibid.*
[357] Löwendahl to Maurice de Saxe and La Tour to d'Argenson, camp before Ostend, 17 August (*Ibid.*).
[358] Löwendahl to Maurice de Saxe and La Tour to d'Argenson, Ostend camp, 18 August. – Beaumont to d'Argenson, camp before Ostend, 17 August (Corresp. A.F.).

Henry Pichat

According to the dispositions taken by the maréchal de camp of the day, at 11:00 p.m., on the 22nd, upon a signal given by several bombs, four companies of grenadiers (supported by a support element which was not to move out except in case of need) and 250 workers marched into the salient, along the sea, to attack the left face of the work. A similar attack was launched against the right flank of the work. The bastion was carried and the French later fell on the covered road's guard. The grenadiers carried it while the workers rapidly prepared the defenses necessary so that the soldiers could hold the captured work.[360] Upon the signal the two attacks moved forward. From the first instance, one could not count on surprising the garrison. They opened a violent fire on the left attack, where grenadiers had to move over about 100 toises of open ground. Despite the darkness of the night, the French climbed a spur crowned with a palisade, reaching the crest of the covered road, and jumped into it. Instead of taking refuge in the works that the workers had dug, the grenadiers, carried away by their ardor, advanced in to the covered road on the guard corps. The garrison greeted them with a murderous shower of bombs that they threw on them in profusion.

The right attack was no less murderous. Debouching at 40 toises from the objective, the grenadiers rapidly jumped into the covered road. Carried away, like their comrades, by an equal ardor, they received the same greeting. A very violent fight began in the road where the garrison finished by withdrawing, leaving 400 men on the ground and 80 prisoners.[361] Three hundred thirty-two French soldiers were hors de combat.[362] During the morning of the 23rd, the French became masters of the bastion on which the grenadiers finally found themselves under cover. At 7:00 a.m., de Chanclos requested that he be given two hours to carry away his dead; but at 8:00 a.m., he began to talk about capitulating.[363] The "great effort" had completely

[359] *La Tour to d'Argenson, 20-21 August, and Löwendahl to Maurice de Saxe and d'Argenson, Ostend camp, 21 August, (Ibid.)*

[360] The disposition of the attack on the covered road at the salient angle of the sea bastion by d'Hérouville (Corresp. A.F.). – This maréchal de camp found himself in the trench at 2:00 p.m., on the 22nd to the same hour on the 23rd." (Order of 21-22 August 1745). *Ibid.*)

[361] La Tour to d'Argenson, camp before Ostend, 24 August (Corresp. A.F.).

[362] State of dead and wounded in the attack on the covered road (*Ibid.*). – This document lists 120 dead and 212 wounded, presenting the details by regiment.

[363] Löwendahl to Maurice de Saxe, camp before Ostend, 23 August (Corresp. A.F.).

Maurice De Saxe's 1745 Campaign in Belgium

succeeded. The governor insisted on receiving the honors of war. Already, on the 16th, Löwendahl had solicited and received instructions in light of a possible capitulation so the negotiations were quickly ended. On 27 August, the garrison of Ostend came out of the fortress to be taken to Mons, where it was to arrive on 11 September. The Allies vainly asked that it be sent to Antwerp.[364] The French lost 250 dead and 486 wounded, officers and men, during the siege.[365]

This rapid denouement permitted a quick beginning of the siege operations of Nieuport, which the coming of bad weather demanded rapid attention.

Already, on 19 August, Löwendahl had sent some reconnaissance detachments to examine this fortress. None of the defensive preparations made by de Gibson, Governor of Nieuport, had escaped him. The governor repaired the casemates in the ramparts and withdrew from the outer works, that were easy to capture, the artillery that they contained.[366] He had done everything necessary to send, upon the first alert, a strong detachment to Lombardzyde.[367] He supported his flank on the little Fort of Nieuwendamme and his forces intended to block the road from Ostend, over which Gibson expected Löwendahl to advance. At Gibson's instigation, the magistrates of Nieuport had invited the bourgeois to draw from the region of Furnes all the provisions possible while there was still time.[368] The garrison consisted of 4 battalions, barely 2,000 well supplied men, because the governor had vainly requested a reinforcement of 1,000 men.[369] Otherwise, during the siege of Ostend, Gibson had remained in communication, by sea, with Chanclos, so one could not count on surprising Nieuport. Finally, on 15 August, to absolutely put his place in security, Gibson flooded almost the totality of the Chastellany of Furnes, despite the pleas of the inhabitants of the region that would be ruined by this act. The peasants, however, had "recently greased Gibson's leg" [bribed Gib-

---

[364]Courbuisson, commander at Bruges, reported to d'Argenson, on 30 August, that on the 19th, the garrison of Ostend crossed through Bruges. "The passage," he said, "lasted from 7:00 a.m., to 8:00 p.m.," It was given the food necessary and Courbuisson sent the report to the Minister. This document mentions 3,577 rations. This document alone permits us to approximate the number of defenders of Ostend.
[365]State of dead and wounded during the siege of Ostend (Corresp. A.F.). The details are provided by regiment.
[366]Anonymous report on Nieuport, 10 August (Corresp. A.F.).
[367] *Ibid.*
[368] Anonymous report on Nieuport, 19 August (Corresp. A.F.).
[369] *Ibid.*

son] so that he would not resort to this extremity, if one is to believe Löwendahl.[370]

On 26 August, the latter reported to Maurice de Saxe that he was going to put his troops on the march for Nieuport and that the preliminary dispositions had already been made. Twice Löwendahl had reconnoitered the direct road linking these two fortresses along the channel. Water made this path difficult, as did the low tide.[371] Another route had been prepared by Liffinghe and Furnes. The flooding only permitted an attack on the west front of Nieuport. It also prevented camping troops between that place and Oosidunckercke. As a result, after having reconnoitered and repaired the road that he proposed to follow, Löwendahl prepared to move down it with his troops, artillery, and the train carrying the fascines so painfully constructed during the siege of Ostend. As for the reinforcement of artillery sent by water from Dendermonde, it was to remain in its boats, because Löwendahl did not foresee that he would need it.[372] Beaumont, intendant to the siege corps, received orders to assure with the resources of Bruges the distributions of bread, meat, and rice for four days as well as forage for two days. After the installation of his troops before Nieuport, the region around Furnes would provide the necessary provisions.[373] The mobile hospital detachment installed at Liffinghe would remain there and one would only bring forward the ambulance service from the old camp.[374] Finally, Beaumont received from Courtrai a supplement of 200 horses, and he could requisition, without difficulty, the wagons necessary for his transport from the region around Bruges.

The itinerary chosen to reach Cauxdye and Oostdunckercke by Liffinghe, Slype, Schoore, and Wulpen necessitated the construction of bridges in the two latter localities as well as some work to facilitate the passage over some streams. If it was unlikely that the weak Nieuport garrison would come out to engage the French in their approach, it was necessary to expect to see the garrison cut the dikes

---

[370] Löwendahl to Maurice de Saxe, camp before Nieuport, 30 August and 1 September (Corresp. A.F.).

[371] "I have twice reconnoitered the road by the channel at low tide. I am not very happy. Apparently, a head wind influences there, because it is found there water reaches to the saddles of horses and the width of water that must be passed varies from 500 to 600 paces." Löwendahl to Maurice de Saxe, Ostend Camp, 26 August (Corresp. A.F.).

[372] (*Ibid.*)

[373] Beaumont to d'Argenson, camp under Ostend, 26, 27, and 28 August. (*Ibid.*)

[374] (*Ibid.*)

on the Ostend, Dixmude, or Furnes Canals. The waters would then flood all this region. To prevent this, on 26 August, Lieutenant Colonel de Montghion had left Osten with 500 fusiliers, an artillery officer, and 50 dragoons. He occupied Lombarzyde and Saint-Jooris with the mission of watching the canal and preparing the passage for the artillery.[375] The following day the few Allied troops that occupied Nieuwendamme evacuated it. On the 27th, Lieutenant Colonel de Lugeac left Ostend in his turn with 700 men to occupy Ramscapelle. He was to link his reconnaissances with those of Montghion and to guard the bridges at Schoore and Wulpen.[376]

Finally, on the 28th, Löwendahl sent his troops forward in a single column.[377] He only brought part of his artillery with him. The rest, escorted by the Royal Artillery, would not join him until the 31st. The camping parties left at 6:00 a.m., with four battalions of the Grenadiers Royaux. The baggage followed this advance guard at 7:00 a.m., and the troops marched out at 10:00 a.m. At the same time, La Marck left Lissemoris and camped at Mariekercke, not joining Löwendahl until the following day. The battalion of Saint-Brieuc Militia remained to guard the mobile hospital and the five squadrons of the Beauffremont Dragoons waited at Mariekercke for the arrival of Louis XV, who had gone to visit Ostend.

On the 28th, Löwendahl and his troops camped between Slype and the Tempelhof commandery. On the 29th, they were at Schoore.[378] The general did not stop at this last location; he pushed as far as Wulpen and Oostdunckercke with four battalions of the Grenadiers Royaux and the Crillon Regiment.

On the 30th, the column joined him. The troops occupied their camp and the following day the cannon arrived.[379] On 1 September, finally, the two detachments of Lugeac and de Montghion joined the main body, leaving only 150 men in their respective ports. The investment was completed. The weather had happily improved. If the rain had persisted the troops could move only with the greatest difficulty in the swampy region of Furnembach.

---

[375] (*Ibid.*)

[376] (*Ibid.*) 28 August.

[377](*Ibid.*) 28 and 30 August. – Löwendahl to Maurice de Saxe, camp before Ostend, 28 August. – Movement order for the troops of von Löwendahl. (Corresp. A.F.).

[378]La Tour to d'Argenson, Oostdunckerke Camp, 30 August. – Beaumont to d'Argenson, camp before Nieuport, 31 August (Corresp. A.F.).

[379]Löwendahl to Maurice de Saxe, camp before Nieuport, 1 September. (*Ibid.*)

The siege operations began immediately. The attack on the chosen front could not be pursued except at two points, one part along the shore in the dunes and the other on a tongue of ground about 300 toises wide. The first direction, called the left attack, had as its objective Fort Viervoet. The second, the right attack, was directed against the bastion from behind which flowed a branch of the Yperle serving as a ditch to the ramparts of the body of the fortress. Between the two attacks there extended an inundation, an important lagoon in the form of a V, about 500 toises on its greatest side. It was indispensable that one capture Viervoet because it was the only means of blocking access to Nieuport from the sea and preventing the Allies from attacking the rear of the right attack.[380]

Operations began simultaneously on both attacks with great activity and everywhere dirt was being moved, from 31 August, to open the first parallel. The garrison had begun firing with 20 cannon and four mortars from the fortress, joined by the four guns and two mortars at Viervoet. At 7:00 a.m., on 1 September, eight mortars on the French left responded to the fort's artillery. The French fire had little effect because of the small surface area of the fort and the lack of skill on the part of the French gunners.[381] A battery of four 16 pdrs was unmasked the following day on this object and produced the most beautiful effect.[382] On the right, the French workers were only 100 toises from the place and the three engineers prepared to bring into action three batteries of six guns and 12 mortars when the attacker on the left saw, in the night of 3/4 September, that the garrison had in part evacuated Viervoet,which was ruined by the French fire. Three companies of grenadiers and three pickets were sent to occupy the fort. The French soldiers entered it by the gorge. Of the 180 defenders who had occupied this work, there remained only a sergeant and 15 Austrians. They loosed a volley, threw down their muskets, and fled so quickly that one could not capture them, except for one that was less alert than the others. Though the Austrians looked as if they would return, the French immediately began their attack on the communications linking Viervoet to Fort de l'Écluse which blocked the Yperle. The first redoubt covering this communication was taken as

[380] (*Ibid.*)
[381] La Tour to d'Argenson, camp before Nieuport, 2 September. – "I am not happy with these bombardiers." [Löwendahl to Maurice de Saxe, camp before Nieuport, 2 September (Corresp. A.F.)],
[382] Löwendahl to Maurice de Saxe and La Tour to d'Argenson, camp before Nieuport, 4 September. (*Ibid.*)

thus the French found themselves halfway to Fort de l'Écluse. Six cannons armed this last work, and an equal number were established in Viervoet to destroy Fort de l'Écluse as soon as possible.

During this time, the work on the two attacks was completed and, during the morning of the 5th, the French artillery was in a state to open fire.[383] On the right the French were 50 toises from the objective and were under the protection of 18 cannon and 22 mortars and were ready to send forward their assault columns. The fall of Fort de l'Écluse, threatened on both sides, was inevitable. De Gibson did not await the attack. At 6:00 a.m., he had the chamade [call for a parley] sounded and raised the white flag over the threatened work.[384]

For several days already, this governor was barraged by the complaints of the inhabitants. The French cannon had ignited numerous fires in Ostend during the siege. The bourgeois of Nieuport, alarmed by this, begged Löwendahl to spare their homes. Löwendahl responded that it was best for them to insist that de Gibson had not flooded the land which ruined it without any profit to the defense of Nieuport.[385] Did the governor cede to the pleas of the bourgeois? Did he fear a renewal of the great effort given by the French grenadiers on the covered road at Ostend? Be it as it may, he asked to capitulate. The French now only had to cross the ditch of the fortress; yet when he was told that his garrison would become prisoners of war, he showed some perplexity. His hesitation did not last long. He took the council of his officers and the terms of the capitulation that he had received at 7:00 a.m., were accepted that evening.[386] The garrison, consisting of 60 officers, 20 sergeants, 1,906 soldiers, and 25 cannons, marched out on 9 September.[387] Escorted by 100 horsemen and 100 infantry, they were led to Dixmude, on the 9th; to Ypres, on the 10th; to Warneton, on the 11th; and to Lille, on the 12th, where they arrived greatly reduced, a number of Austrian deserters having taken service in French free com-

---

[383] Löwendahl to Maurice de Saxe and La Tour to d'Argenson, camp before Nieuport, 4 September (Corresp. A.F.).

[384] Beaumont to d'Argenson, camp before Nieuport, 4 September; Löwendahl to Maurice de Saxe, Nieuport Camp, 5 September (*Ibid.*).

[385] Löwendahl to Maurice de Saxe, Oostdunckercke Camp, 1 September (Corresp. A.F.).

[386]La Tour to d'Argenson, camp before Nieuport, 5 September (*Ibid.*).

[387] State of the garrison of Nieuport, presented by Foulon, Commissaire-Ordonnateur of the corps of von Löwendahl, to Maupassant, Commissioner of War, charged with the conduct of prisoners at Lille (*Ibid.*).

panies. The French had lost 22 dead and wounded during the siege.[388] During the passage of these events, the situation of the Allies did not improve. The landing of Charles Edward in Scotland and his very rapid progress did not permit King George II to send any reinforcements to the Continent. This diversion further augmented the importance of Antwerp. It obliged, in effect, that the English retain this city, which was more precious than ever for the communications with the Pragmatic Army in case the London Court was obliged to recall its troops from Flanders in England. This aggravated the fears of the Allies for their right wing. On 6 August, the Duke of Cumberland did not conceal his anxiety.[389] "The English generals," said von Waldeck, "came to beg von Königsegg to pull the army towards Antwerp, telling him that it was impossible to protect Antwerp and Brussels at the same time, to which the Marshal said to me that he wished to speak. I went to him and I strongly opposed the request of the English generals to point out the advantage of our position in Brussels and the results of abandoning this city, which would cause us to also lose Mons and Namur in this campaign." It appeared that the Allies were discussing what they would abandon to the French. The perpetual temporizer, von Königsegg took a position that satisfied everyone. On the 7th, he extended the right of the Pragmatic Army to a point below Antwerp at the Saint-Bernard Abby and on the side of Brussels he extended the flooding across all the land between the Lacken and Hal Gates. He brought his troops closer to the canal on which he constructed an entrenchment lined with redoubts.[390] One can say that this resolution was the only one that von Königsegg could take in this situation. In acting differently, he risked provoking the separation of the Allies, an eventuality that the French spies had frequently suggested was quite possible.

The following day four Austrian battalions, which had come from Namur, reinforced the Pragmatic Army.[391] These reinforcements appeared to bring to the Allies a worsening of morale. The Allies prepared a battlefield before their front as well as the roads necessary to move their artillery. The English and Dutch evacuated their wounded, the first towards Antwerp and the second towards Mons and Namur. This renewal of activity reached its paroxysm in a conference

---

[388] State of the dead and wounded during the siege of Nieuport (*Ibid.*).
[389] Waldeck, 6 August.
[390] Waldeck, 7 August.
[391] (*Ibid.*) 8 August.

Maurice De Saxe's 1745 Campaign in Belgium

held on the 11th with Königsegg.[392] The presence of Clermont-Gallerande at Chièvres, not far from Ath, disconcerted Prince von Waldeck, who proposed nothing less than throwing out the French troops. It is necessary to say that at this same moment the French occupied themselves with making fascines and that sufficed to raise Waldeck's suspicions.[393] The Prince sought to capture the French detachment between a detachment drawn, for this occasion, from the left wing of the Pragmatic Army and parties that would come out of Mons.[394] Happily for Clermont-Gallerande, the proposition of the Republican general was not accepted. However, the Prince appeared particularly in the grip of the general crisis of activity existing among the Allies. He resolved to send a reconnaissance towards Assche, a region "where," he said, "I believe that the enemy [French] army is weakened by its various detachments."[395] This reconnaissance appears to have been premature, as it did not appear that the French army would soon weaken itself. The situation on 13 August established that at this moment the troops in line at the Alost camp, including the Grassins, totaled 79 battalions and 125 squadrons. The 24 battalions and 5 squadrons with Löwendahl were about to return and in the conquered places the French established garrisons mostly formed from militia.[396] One can suppose that, perhaps, the Prince hoped to put into execution by a different means, the project that he had earlier suggested, and which had no other goal than drawing the troops of Clermont-Gallerande away from Ath by whatever means possible.

During the morning of 12 August, he left his camp with 400 horsemen, 400 infantry, and a free company, moving on Assche. After crossing through Brussels, he moved down the highway when the inhabitants of the suburbs informed him of the presence of about 2,000 French in the vicinity of Assch.[397] He continued to advance, giving little credence to this information.

The intelligence, however, was not completely incorrect. The day before Maréchal de camp Beausobre left Ninove with about 1,000 men, uhlans, and hussars, pushing a reconnaissance as far as Steenussel.

[392] (*Ibid.*) 11 August.
[393] R.A.
[394] Waldeck, 11 August.
[395] Waldeck, 11 and 12 August.
[396] The French had, in Bruges, 2 battalions; at Audenarde, 2; at Tournai, 6, and at Ghent, 6, with 12 squadrons. Among this group of troops, only 4 battalions and the 12 squadrons were not militia.
[397] R.A.

Henry Pichat

There, he made contact with the patrols sent daily into the area of Harcourt, throughout all of the siege of Dendermonde. [398] Seeing nothing suspicious, Beausobre returned, during the morning of the 12th, down the same road as the day before, misunderstanding the orders given by the maréchal de camp for the conduct of reconnaissances of this nature. [399]

Elsewhere on that same morning, two other parties left the right of the French camp to execute the daily security patrols directed by Maurice. Three hundred volunteers, with Méric formed the first and 600 horse of the King's Household, commanded by Lieutenant General d'Estrées, led the second. [400]

Von Waldeck arrived a half mile from Assche without encountering anyone. He then put his infantry in an ambush under the cover of some hedges that lined the road to Brussels and pushed his hussars forward to reconnoiter the village. [401] He followed them in person with his cavalry and a free company. [402] Some minutes later after Beausobre arrived unexpectedly at the forking of the road to Steenussel and the highway, that is to say about 500 meters to the east of Assch, he moved on the village. According to a friend of the Count d'Estrées, Beausobre took no precaution to "search" the village. Brézé relates, to the contrary, that the maréchal de camp sent his grenadiers to do this and that he detached his uhlans and hussars "beyond, to the right and left" of the village. [403] The phases of the combat that then resulted lets one suppose that Brézé's account of the steps Beausobre had taken are correct.

Whatever may be, the French entered the village and were greeted by a fusillade by the Allied scouts. They succeeded in driving back the Allies to the edge of Assche facing Brussels and there "they engaged in a very sharp fight that lasted for some time." [404] To finish, Beausobre attempted to envelop the Allied troops with his cavalry. This maneuver decided Waldeck to withdraw on his infantry which had come out of its ambush and arrived to the rescue. [405] The French soldiers pursued the Prince, but were greeted by a discharge by the

---

[398] Waldec, 12 August – Report to the States General on the action at Assche (A.L.H.).
[399] R.A.
[400] Brézé.
[401] Waldeck, 12 August. – Report to the States, *loc. cit.*
[402] (*Ibid.*).
[403] Brézé.
[404] Waldeck, 12 August. – Report to the States, *loc. cit.*
[405] Brézé, siege of Dendermonde.

Allied infantry, then charged by the English cavalry, they were driven back to Assche.

Otherwise, for some time already Beausobre noted, on his left, that is to say, towards the north, a force of cavalry and infantry. He thought it to be Allied forces. Not wishing to have to undertake a battle to the east and to the west of the village, he ordered his troops to withdraw behind Assche and to leave for the Alost camp. However, this was not an Allied reinforcement that was arriving, but the Count d'Estrées, Méric, and their soldiers. They were recognized as friends and the lieutenant general took command. Brought up to date on the situation, he learned that the forces before him formed the advance guard of a considerable corps and that von Waldeck led it in person. Without seeking to verify such important information with one of Beausobre's light horsemen, d'Estrées confirmed the orders to retreat. [406]

At this moment Méric arrived unexpectedly. Arriving with his infantry on the terrain of the action less quickly than d'Estrées, Méric had seen Waldeck form a rearguard. He asked for the honor of charging the enemy. [407]

Méric was not fooled. Waldeck had seen in d'Estrées' arrival a confirmation of the information received at his departure from Brussels and, as a result, he decided to retreat. A rearguard, given to Cornabé, his faithful adjudant, received the mission of covering Waldeck's withdrawal. Beginning his movement, Waldeck could not divine at the same time that Count d'Estrées was, like him, withdrawing. [408] Other circumstances will show later with which facility this lieutenant general created unfortunate certitudes, which infallibly followed him. Méric arrived in time to prevent d'Estrées from covering himself with ridicule.

Having received the requested authorization, the partisan took the road with his volunteers, supported by some hussars. Cornabé presented a good face. He overcame "obstacle after obstacle" without allowing himself to be turned, even though he was attacked "with all vivacity possible." [409] After an hour, at Zellichk, the pursuit ended. [410] Pushed further, it would become dangerous because of the proximity

[406] Report to the States, *loc. cit.*
[407] Brézé, siege of Dendermonde.
[408] Report to the States, *loc. cit.* – Waldeck, 12 August.
[409] Report to the States, *loc. cit.* – Waldeck, 12 August.
[410] Brézé, siege of Dencermond.

of Allied posts. The French then returned to their camp. The brave Méric had been wounded, and the French suffered 4 dead, 38 wounded, and 16 prisoners.[411] The Allies lost 8 dead, 28 wounded, and 16 prisoners. The action would have no other consequences. In his *Journal*, Schlippenbach gave to this reconnaissance action its true character. "Von Waldeck had the desire to go for a promenade accompanied by a good escort with which he risked going near Assche, where he met a greater number of enemies than he anticipated." One would have a hard time finding a more impartial and concise chronicler that von Schlippenbach.

On 16 August there appeared the "project for the disposition of the end of the campaign." Maurice estimated that the sieges of Ostend and Nieuport would extend to the end of September and end the campaign.[412] This hypothesis permitted him to anticipate that the first days of October would be spent by the French troops in their winter quarters prepared on the left bank of the Escaut. Under these conditions, the army, "whose conservation appeared preferable to everything," would be installed at the beginning of the bad weather.[413]

In awaiting the moment to enter winter quarters, it consumed, in place, the provisions drawn from the land between the Dender and the Senne. This stationing obliged, at the same time, to not separate their troops prematurely.

As a result, Maréchal de Saxe raised camp, on 17 August, to install himself towards Lippeloo. The defenses of Dendermonde were destroyed first and one lived with the greatest economy in the quadrangle between Dendermonde, Alost, Brussels, and Rupelmonde. Clermont-Gallerande moved on Hal or Enghien so as to execute Séchelles' orders and to assure the resupply of his troops. Séchelles formed three magazines. The first was at Ninove and contained all the food drawn from the region of this small fortress situated in a very broken terrain for an army as large as the French to camp and live there comfortably. The convoys of provisions protected by the hussars and Grassins converged at Lessins. The transports of this place at Ninove were effected by means of boats requisitioned on the Dender and the Escaut. The second magazine was formed at Alost with everything that could be found between Enghien and the right of the Dender. As for the third

---

[411]State of the dead and wounded in the detachment of de Méric at the engagement at Assche (Corresp. A.F.).

[412] De Vault, p. 63

[413] (*Ibid.*).

depot, it was constructed at Ghent with the supplies furnished by the Waes region. Detachments of the army camped at Lippeloo protected the transports in this region. "By this means," concluded the Marshal, "I count on assuring us with abundant supplies for the rest of the campaign and to take it from the enemy."[414] The army would then return, the magazines filled, and camp between Alost and Ninove.

At the beginning of its execution, this project underwent a modification. The French hesitated to destroy the defenses of Dendermonde. The Marshal had thought that it was better to dismantle the fortress than to leave a garrison in it. In these conditions the Allies could not re-occupy Dendermonde and install themselves close to the French winter quarters. Before beginning this work, a report was requested from the engineers. It was provided on 20 August.[415] This report showed that rendering Dendermonde indefensible, it sufficed to raze the fortifications, the wall, and the redoubts. If, to the contrary, one wished to put it into a state of defense, the engineers estimated it would be necessary to employ 800 pioneers for 30 days and spend 9,000 livres. Finally, one could destroy the totality of the sluices. If one took this step, the region would find itself, according to the men of the art, entirely lost, because "there would remain no resource for the control of the waters." This consideration decided the Marshal to keep the water recourses in place "less with an eye to protecting them," said de Brézé, "than the sad apprehension of destroying the land."[416] Two battalions were left there.

On the fixed day, the army left for the Lippeloo camp or Mélis. In light of assuring the security of this march, Chevreuse and four regiments of dragoons had occupied Opwick, on the 16th. On the 17th, the camping detachments assembled at 4:00 a.m., leaving immediately, escorted by 12 companies of grenadiers. The troops followed in eight columns. The right wing and the 2nd column, formed exclusively of cavalry, and the 3rd, 4th, and 5th of infantry. The 6th was for the mobile hospital, the artillery, and the pontoons; the 7th contained the cavalry of the left of the two lines of the camp, and the 8th, finally, of the Carabiniers, the Guards, the King's baggage and that of the headquarters. Each column brought it in its suite the heavy baggage

---

[414] Project of dispositions for the end of the campaign, 16 August (Corresp. A.F.).

[415] Report on Dendermonde, 20 August (*Ibid.*). – This interesting document is signed d'Aumale, d'Oyré, Courdommer, Filley, d'Aboville, and Brest de Lépinoy, captain of miners.

[416] Brézé, 24 August

Henry Pichat

accompanied by some escorts and ordinary rearguard.[417] The King and his suite left Alost.[418] Following de Brézé, the movement order ended with a disposition that appeared to be inspired by the sight of some new Royal fantasy troubling the march.[419]

With an eye towards covering the left wing and the establishment of the camp, Grassin sent parties to block the interval between the Escaut, the Lippeloo Stream, and that of the Londerzeel. He also masked the passages of this last watercourse.

The day before the d'Auvergne Brigade had been detached. The 1st Battalion, d'Auvergne was sent to Alost. The 2nd and that of the Royal la Marine reinforced a post of Grassins in Afflighem that was there and remained there. The 3rd Battalion moved to Ninove, under the command of Beausobre, previously installed in this post with his uhlans and hussars. These troops were to assure the security of the future magazines.[420]

At the same time Clermont-Gallerande received orders to leave Chièvres and move to Enghien.[421] "The position that we are going to take tomorrow before the enemy," the Marshal wrote to him, "will put you in a position to make a movement on Enghien, if the issue of provisions obliges me to inform you that the move cannot occur until after they have been arranged."[422] De Clermont-Gallerande marched out only when Commissioner Crancé, sent into the Brabant as a delegate to Séchelles, had advised that he was in a position to send his requirements. Otherwise, Clermont-Gallerande remained free in his movements. "I will not obstruct you on that which you may judge appropriate to do," added Maurice, "their suitability depending on many circumstances of which nobody can better judge than those who are on the scene. It is not necessary either that you scrupulously restrict yourselves to a certain position. I leave you the master of that decision; you are there to produce jealousy and as soon as you reach that point, you leave nothing to desire."

---

[417]Order for 17 August (Corresp. A.F).

[418] Louis XV arrived at Alost on 4 August.

[419] "By this route (Alost, the mill at Wiese and Lippeloo) the King crossed several columns; but it was better to follow the same road as they did, which without this, would be inevitable and if His Majesty did not wish to leave before 9:00 a.m.; truly the columns would have passed and the King would have found nothing to obstruct his march." (Brézé, March Order for 17 August).

[420] Auvergne (3 battalions) and Royal la Marine (1 battalion) formed the brigade.

[421] Maurice de Saxe to Clermont-Gallerande, Alost, 16 August (Corresp. A.F.).

[422] (*Ibid.*).

When he received these orders, Clermont-Gallerande waited for four days for bread sent by the Condé Fortress. The Marshal's courier arrived so late that this provision could not be cancelled in time, such that Clermont-Gallerande did not move on Enghien until 21 August.

The French Army established itself in a new camp extending from Merchtem to Lipeloo by Stennussel in two lines facing to the east. A stream, "good by the heights of its banks" ran across the front and the four regiments of dragoons camped below Merchtem covered the right.[423] The supply operations prescribed by the project for the end of the campaign began immediately.

If the Allies did not stop them in them in the regions neighboring the camps, they would seek to pillage the convoys directed on Lessines in the areas where the surveillance of the French hussars was limited. He sought, thus, to exercise reprisals against the population that submitted to Séchelles' orders. In the vicinity of Mons and Ath the Austrians repeated their orders to provide nothing to the French "under the pain of corporal punishment."[424] In order to conform to the instructions of the intendant, Crancé had directed 60,000 bundles of forage to the French magazines. He had sent his requests to the Chatelleines of Ath, Mons, Binch, and Braine-le-Comte; however, the Allied hussars carried off 60 wagons leaving Quiévrain. Besides, the French spies reported also that the Allies had not received the large quantities of provisions that flowed into Brussel, their commissioner of supply officers having the last markets on 25 October. Certain people had spontaneously offered this subsistence as a reduction to the impositions still due to the Queen for the year 1745. As the provisions due were much below the value of this offer, there was surely, at the bottom of this eagerness to provide supplies, only a desire to deprive the French of all the supplies transported to Brussels.[425] At the instigation of these same individuals, the peasants around Assch buried at night all the grain gathered during the day and then clandestinely transported it to the Allies. "There was soon more in Brussels than in the fields," wrote the Marshal.[426] The population complained of this pressure from the Allied agents who exposed them to the French rig-

---

[423] Maurice de Saxe to Clermont-Gallerande, Alost, 16 August (Corresp. A.F.).

[424] Pons, 18 August.

[425]Thirion to d'Argenson, Valenciennes, 13 and 14 August (Corresp. A.F.).

[426] Anonymous letter from Brussels, 26 August (Corresp. A.F.).

ors.[427] On his side, Clermont-Gallerande did not bring all the energy necessary to execute the orders of the intendant.[428]

To bring an end to these maneuvers, the Marshal resolved to deal ruthlessly with the Allied hussars and free troops whose expeditions left from Grimberghen, Kockelberg, or Brussels itself. Grimberghen, in particular, appeared to be as paradise for the Allied pillagers among whom figured a number of French deserters. They lived there in abundance. "I doubt that Grimberghen would get over it," wrote the French correspondent in Brussels to the marshal.[429]

Maurice had already attempted to chastise them. On the morning of 8 August some Grassins and some infantry failed to capture the free company from Grimberghen. This latter barely had time to seek refuge in the castle from where a reinforcement saved it.[430] A new effort was attempted in the afternoon of the same day and it also failed. On the 20th, a third fruitless attempt gave rise to a violent battle. Grassin sent from Afflighem three detachments of 30 men each, by different routes, all of which was supported by a company of grenadiers, a total of 150 men.[431] One of the groups was so imprudent as to let itself be seen towards Kockelburg and immediately the pillagers took chase. The Grassins fell back on the grenadier company in ambush in Zellick. By chance Waldeck was making an inspection of these posts at this time. He sent out some troops from the vicinity to which he joined the dragoons of his escort and this force closed in on the 150 French soldiers in Zellick. For two hours the French maintained a stubborn battle, but only 20-30 were able to escape, the rest being killed or captured. The following day the Alies reinforced Grimberghen with an Austrian free company.[432] It became urgent to act immediately.

---

[427] Anonymous letter from Brussels, 27 August (Corresp. A.F.).

[428] "The peasants furnished nothing more than peas and vegetables of an inferior quality. As to sheaves of corn, they had great need to be beaten before they were brought forward. They complained. Clermont-Gallerande responded that the peasants suffered thus from war because the troops contented themselves with what was delivered to them."

[429] "Grimberghen furnished daily 400 pounds of meat, 860 of bread, 600 pots of beer and the abbey provided for the officers. I plan on going there tomorrow." [Anonymous letter from Brussels, 26 August (Corresp. A.F.)].

[430] Waldeck, 8 August.

[431] Scovaud (Major of the Grassins) to d'Argenson, Abbey of Afflighem, 21 August (Corresp. A.F.).

[432] Waldeck, 20 August.

Maurice De Saxe's 1745 Campaign in Belgium

During the evening of the 22nd, an expedition left the Lippeloo camp, commanded by Lieutenant General Le Danois, assisted by three maréchaux de camp, and contained 19 battalions, 24 companies of grenadiers, 120 horse, and some cannon.[433]

After a night march, at 5:00 a.m., on the 23rd, the French advance guard, commanded by de Lorges (12 companies of grenadiers, 12 piquets, 100 carabiniers, 4 cannons) occupied the Grimberghen heights from where they found an important Allied post at Vilvorde and a bridge over the canal. The advance guard deployed so as to stop everything that came out of the Pragmatic Army's main line on this side. Meanwhile, de Souvré, who followed with 12 companies of grenadiers, 12 piquets, and four cannons advanced resolutely on Grimberghen. Finally, Le Danois, in the rear of the column, established himself between Beyghem and Wolverthem in two lines, the cavalry in the first and the infantry in the rear, to block the plain between these villages.[434]

An Allied party occupied the two castles at Grimberghen. The rest held themselves in the village from where about 400 men came out and moved boldly against de Lorges as soon as he appeared. Greeted by cannon shots, the Allies re-entered Grimberghen at the same time as Souvré summoned the two castles to surrender.[435] The smaller of the two, defended by Hanoverians, surrendered immediately. The other, occupied by the English and French deserters resisted. De Souvré had only Swedish guns[436], so he moved closer to make a breach in the walls.[437] At this moment about 2,000 Allied troops, with artillery appeared on the plain of the canal to the south of Vilvorde. The Duke of Cumberland was coming in person on his own initiative with reinforcements. Drawn by the sound of combat, he had crossed the canal and then moved to cross the Grimberghen Stream.[438] The surprise checked the action. Le Danois was going to be obliged to give, in his turn, a fight with all his troops that he now found engaged. Soon the situation of the Duke of Cumberland exposed him to being driven into the stream, which he could only cross slowly. Le Danois

---

[433] Brézé, 24 August.

[434] (*Ibid.*).

[435] Waldeck, 23 August, and Waldeck's report to the States on the engagement at Grimberghen (A.L.H.).

[436] Translator: "Swedish guns" were light regimental guns being either 3pdrs or 4pdrs.

[437] R.A.

[438] Waldeck, 23 August.

gave the order to his lieutenants to withdraw on him and to retreat by Londerzeel. He was pursued only weakly.

The failed attack on Grimberghen did not reduce the ardor of the Allied pillagers. On 31 August, they captured another 200 wagons destined for the magazines at Ninove, which they sought, unsuccessfully, to burn. It appeared that the Allies wished to extend the field of their exactions, because they sent parties to Maubeuge and Charleroi.[439] Despite everything, on 1 September the French magazines on the Dender contained the provisions necessary to feed the army to the 25th of the same month. As a result, Maréchal de Saxe decided to begin the breakthrough marches. Besides, Ostend was captured. The fall of Nieuport was imminent and the sandy and little fertile region where the army was camped was nearly exhausted.[440] However, having no reason to hasten, Maurice resolved to, in two marches, arrive on the left bank of the Dender between Alost and Ninove. By the first movement he proposed to cross the river and to camp between Alost and Dendermonde. He then closed up the army to establish it between Alost and Ninove. Thus, it would find itself placed in the proximity of the Escaut bridges when the moment would come to move to winter quarters on the left bank of the river.

On 7 September, the French Army began its movements, and its department furnished a new occasion to trick the Allies. To suppose that the Allied generals were unaware of the French projects, of the recent departure of the King for Versailles, was sufficient to calm their suspicions.[441] The Marshal learned, with certainty, that on 2 September, the Allies had drawn from their right 15 squadrons and camped at Vilvorde, "to provide word when the French rearguard left Lippeloo."[442] Suddenly, on 4 September, Waldeck heard that the French engineers were laying out roads in the vicinity of Dilighem. He immediately sent out a reconnaissance force.[443] This force pushed to Meysse and returned by Grimberghen without encountering any workers. It found only the trace of the indicated roads.[444] This observation caused the Allies to predict the approach of the French towards

[439]Waldeck, 28 and 31 August. – These troops would commit such depredations that the Marshal was soon obliged to take significant measures.

[440] Maurice de Saxe to d'Argenson, Mélis camp, 6 September (Corresp. A.F.).

[441] Louis XV and the Dauphin left the army on 1 September to visit Ghent, Ostend, and then to return to Versailles.

[442]Waldeck, 2 September.

[443] *Ibid.* 4 September

[444] "It was possibly true that one had laid out these roads." (*Ibid.*).

the canal rather than moving on the side of Dender. The Allies reinforced their camp at Vilvorde with 1,000 men, but they did not repeat
their reconnaissance. They had not doubted for a minute that these
preparations for a march forward were only a feint like those of July
intended to cause the Allies to believe there would be a French attack.
At noon, on the 7th, the Allies saw that they had been fooled again, and
news spread throughout the Allied camp that the French Army was
marching towards the Dender.[445] "I shall send Cornabé to observe the
enemy's march," said von Waldeck. "My adjutant was at Assche; but
the enemy had already passed. Our troops were on the march from the
early morning."

To lighten the columns, the heavy baggage had left on the
6th, with the pontoons, the Treasury, the hospital, and the artillery, except for 40 Swedish guns.[446] All this train was parked below Alost
where it awaited the army. Three detachments of Grassins, each with
30 mounted arquebusiers and 30-foot arquebusiers, furnished by the
Afflighem post, covered this march. They occupied the villages of
Moorsel and Masenzel as well as the road to Brussels, on the 6th, until
5 p.m. and then returned to Afflighem.[447]

The army marched out on the 7th. At daybreak, the camping detachments moved out followed by the baggage in four columns.
This last convoy was strongly escorted and reached the future camp
before the arrival of the troops. The troops marched in six columns
"the left of the camp formed the right of the march and the troops were
in reversed columns."[448] The wings were formed exclusively of cavalry. In the 2nd and 5th columns the formation was mixed with cavalry in
the lead. The 3rd and 4th columns were formed of solely infantry. The
four center columns each brought in their tail five Swedish guns between the two last brigades. Finally, strong rearguards provided with
20 cannon covered the three columns. The commandant of Afflighem
renewed, on the 7th, the same security service as he had the previous
day and everything moved in the greatest order. The French used the
road network as ordinary and at 2:00 p.m., the army was encamped.[449]

On the 8th, the army closed up between Alost and Ninove. The
troops formed in three columns so they could execute a light move-

[445]Waldeck, 5-6 September

[446] The army had, among its light cannon, only 50 Swedish 4pdr cannon.

[447] March order of 7 September (Corresp. A.F.).

[448] (*Ibid.*).

[449] Crémilles (maréchal général des logis de l'armée). to d'Argenson, Camp at Alost,
7 September (Corresp. A.F.).

ment. D'Harcourt remained under Dendermonde with the Carabiniers and Royal-Roussilon Regiments formed a total of 18 squadrons. The post at Afflighem, now useless, was withdrawn such that the French now had more troops than Clermont-Gallerande on the right of the Dender.

The army began consuming the supplies gathered in the magazines at Alost and Ninove. It then regained the quarters that de Saxe had ordered formed.[450] Already requests to be employed flowed out during the winter because the campaign appeared to be ended. At Versailles, they were also convinced that the operations were completed.[451]

With the beginning of the bad season, the military customs of the time, and the dispositions of the Marshal seemed to announce the separation of the army. The court had, however, taken the decision which would extend the campaign for yet another month.

We have already seen that Löwendahl, du Chayla, Clermont-Gallerande, and Beausobre as well as some other officers, and even von Waldeck himself, have drawn sharp criticism. The engagement at Grimberghen furnished to the friend of the Count d'Estrées the occasion to criticize Maréchal de Saxe in his turn. The importance of the detachment given to Le Danois and also the almost complete inaction in which the army fell during the month of August gave this action an importance that it otherwise did not deserve. The anonymous chronicler took the occasion to echo a particularly singular and spiteful assessment:

"The valor of M. le Danois leaves no doubt that he had received the order to withdraw. It obliges one to believe what a general officer among his friends told me: to wit that it was a cutting remark between generals that induced him to prove to M. de Waldeck that he would take from him a post about which this general had publicly said that the French would not dare to attack. It was better to drop this remark rather than undertake to punish it without having first foreseen its consequences. You can say that one shamefully lost the challenge and the finest game in the world."

---

[450] Emplacements of the militia battalions in Flanders, Hainaut, Artois, and the Champagne frontier, 23 August. – State of 22 battalions that were drawn from the Army of Flanders, which one proposes placing on the frontier of the Champagne, the Meuse, and the region of Metz during the winter of 1745-46. – Troops of the Army of the Rhine, which one can place in Alsace, in the Évêchés and Lorraine during the winter of 1745-46, 24 August (Corresp. A.F.).

[451] Waldeck, 8 September

Maurice De Saxe's 1745 Campaign in Belgium

The conduct of operations pursued during the months of August and September inspired the same contemporary to criticisms of a more general order and also more strident than the previous criticisms. They presented, in effect, in their totality, a more detailed and characteristic discussion of the Marshal's tactics and they are, as a result, of great interest.

One will see that much closer to this phase of the campaign the general enthusiasm that had arisen at the end of July was not renewed. If one is to believe this chronicler, the commander did not know how to draw all the results he could have expected from Löwendahl's rapid and successful march on the Escaut as well as the bad material and moral situation of the Allies as they sought refuge behind the Brussels Canal. The chronicler exposes the plan of operations that it would have been preferable to follow, according to him at least, from the end of July. He accumulated favorable arguments for his thesis with much insistence and obligingness. His unconcealed admiration for the Count d'Estrées and his relatives, which he does not conceal, with the other generals, allows us to believe that he gives, with his personal impressions, an echo of the opinions and sentiments of his friends. It is appropriate to cite the complete text of the chronicler.

The author appears to initially regret that the army had not camped below Alost, on 8 July, and then that it had not besieged Dendermonde at the same time as Audenarde. He also deplores that the French had not taken Antwerp. It is, finally, rather difficult to distinguish exactly to whom to address the reproaches of the writer. The preponderance of the opinions of Maréchal de Saxe in the council do not permit any doubt about the personality of the individual responsible for the capital faults enumerated by the anonymous critic.

"The capture of Ghent," he wrote, "had two effects, both of which are normal in war, one being to discourage the enemy by the loss of his magazines, his reputation, and his communications with England, which caused him to abandon the Dender and move closer to Brussels where, still not thinking itself safe, it recrossed the Senne to take up position behind it. It also executed some movements, before going there, that convinces us of its discouragement and its irresolution which were absolutely the greatest advantages that one can offer his enemy. The second effect was to occupy so strongly those who decided in the council, by the joy that it caused in our [the French] army, that the great maxim of Caesar could be forgotten and that there remained an infinity of things to do, but it was thought that everything had been done and that they were so occupied with the siege of Audenarde that nothing else could be done. I have been assured of this by several people who received directions to undertake nothing more in the field. Be it that this opinion initially prevailed, be it that no one imagined anything greater, we were solely occupied with the siege and the precipitous retreat of the enemy, during a time when it was not advisable to hope for it; it was not more harmful to him than if he had awaited the news of our first march forward to be determined there.

"This land is like that of Ghent; all the camps that one can take from that city beyond the Dender to Alost are equally commodious by the subsistence that one can draw from France and because the forage there is also good. The camp at Alost covers both the siege of Audenarde as well as the Borst camp, above all in regard to an army, whose capital is at Antwerp and which is placed behind the Senne and the Vilvorde Canal. Now Audenarde remains in our rear without hope of any succor, requiring only 20 battalions to execute the siege. It took no more and one had placed eight in Ghent and Bruges. The army at Alost covered the two posts and assured as much more the position at Ghent. There remained an army of 84 battalions, all the cavalry, and the dragoons. All that had come from the Rhine were not fatigued because of any siege. Despite all this, one did not believe we could undertake the siege of Audenarde at the same time, although the army could enter the trenches at Dendermonde. One prefers to give the enemy the time to resupply this place than to run the risk, one says, of suffering a check in undertaking too many things. If the enemy had guarded the Dender, one would have reasoned correctly. One would have been more correct in not forcing them in their camp before having completely assured the rear by the capture of Audenarde. However, the enemy having abandoned that part, our position at Alost became stronger. Then, what maneuver could they have executed to disturb us? One would have been most embarrassed at Borst as at Alost where one would have been obliged, in case of an attack, to cover Ghent and Audenarde by two separate corps; the enemy could then undertake on Ghent by the left of the Escaut and at the same time amuse us on the right bank. None of these reasons having prevailed, the army remained at the Borst camp until 28 July, without making the least preparations for another siege. I have touched, thus, one of the most essential errors that one made by the delay of the siege of Dendermonde. One further aggravated the defence, I know not for what reason, until 8 or 9 August. One opened the trench on the 9th, since the fortress surrendered on the 12th, after three days of attack, and 12 or 14 hours of cannonade. That which one had not thought to do at the beginning appeared very easy. Audenarde was of too great a consequence not to have been taken before attending to Dendermonde; however, Ostend was not the same and one did not think that one could undertake two sieges at the same time let alone the two that I have cited. I believe that the movement then excited by England was one of the principal causes of the surrender of Ostend and Nieuport which could no longer count on relief. Those of whom could wait were then consequently to stand against the party of Prince Edward Stuart who at this time started to appear formidable in Scotland. In assembling all the reasons and the facilities to make these two conquests, I believe that one can boldly conclude that it was a capital error to linger at these two places for the time that they did.

"The reasons to attack Ostend were the reputation of the King's arms, the necessity to occupy a post by which the English could land forces on this side of the sea, and the ability to take it during the low tides of August and September, as the rest of the year the sea flooded the vicinity so much that one could not think of digging trenches. To report faithfully the contrary opinion, it is necessary to respond that these reasons, of which the first had been the reputation of the King's arms. This is, according to me, the worst reason. This reputation had made a good enough effect at another time and the utility is almost always preferable concerning conquest. The disturbances that we caused in the two fortresses in our rear was much less founded than it would have been had England been in a state to make a diversion by there; it would not have succeeded because of the difficulty of the land around Ostend and Nieuport, which would provide neither the wood, nor the water, nor the forage necessary to support a corps sufficiently large to disturb us.

"Beyond this the kingdom did not think of it, this is what one saw in the plan of which one was instructed, and that began to hatch for now, it was only good to wait for the success of the enterprise to proportion to which one delayed it. This gave time for Prince Edward to form his party. The ease that could be glimpsed in respect of the sea still subsisted in the month of September, thus one could procrastinate until that time and this is why I would like to postpone it."

Henry Pichat

"The principal objective of the campaign in Flanders was to draw out the root of the war in weakening the English to prepare the coming of the revolution and to deprive the Queen of Hungary of this resource without which she could not sustain the war.

"The second objective should have been, along the same lines, to intimidate Holland to engage it to accept an exact neutrality or at least to make no more intimate alliance with our enemies. This was, unquestionably, to attach itself to the root of evil in fully succeeding and working usefully for peace.

"The siege of Ostend was perfectly distant from these two objects. It did not contribute to the revolution in England and the English army overseas was less free to run to the assistance of the Elector of Hanover. It had moved by Antwerp and could not move otherwise, in light of the fact that we were absolute masters of the Escaut.

"The States General did not conceive nor could they conceive any concern for the reduction of this fortress. They saw this with pleasure as it gave them time to put Antwerp into a state of defense, it being the only barrier that remained between us and their state. Besides, I believe that it is easy to prove that we would not think of Ostend until after we had reduced this key to Holland. However, one tells me, the season was too advanced to undertake this. It was not as if the great and deep ideas governed our minds again after the capture of Ghent while forming the siege of Dendermonde at the same time as the one of Audenarde; this place would not have held out one minute longer. We would have been masters of it from 21 July. Then, the army, with 110 battalions and a force of cavalry equal to that of Darius by number, would have been in a state and free to maneuver. Is it probable that the Allies, to whom there remained no more than 35,000 men at most after they had garrisoned Ostend, Nieuport, Dendermonde, Audenarde, Mons, Alost, Namur, and Charleroi, would have risked battle even though covered by the Vilvorde Canal and the Senne?

"They surely would not have and the arrangements that they took after the capture of Dendermonde when the army marched against them, show that they were not of this disposition, even though they had already the leisure to fortify the front of their camp, as they knew part of the army occupied Ostend and 26 squadrons had left for the Moselle.

"They had prepared magazines and the lines behind the Rupel where they wished to march if they were attacked. Otherwise, when they waited, what could one better desire than to give battle with so many advantages on the side of numbers and with troops filled with their successes against troops that they had already beaten twice before and which were filled with discouragement and the distrust of each other to the point where they had come to blows more than once in their camp?

"There are some people who would destroy this reasoning thus: but if you had marched on Antwerp, the Allies, while you attacked the Vilvorde Canal, would have put 20,000 men into Antwerp to fall on Ghent, by the left of the Escaut. This reasoning destroys itself. What would have become of the 15,000 men remaining in defense of the canal? Would they sacrifice this half of the army to execute a mission and leave the rest of Flanders and Holland fall prey to the victor?

"What would the 20,000 men in Ghent have done between Audenarde, Dendermonde, and the army? One cannot imagine! Besides, was one not in a state in marching on Vilvored to put 20 battalions and 30 squadrons under Dendermonde on the left bank of the Escaut? Who would have dared exposing their flank to a corps of this type, as well posted and within range to be reinforced hour by hour without the knowledge of the enemy?

"I have never doubted, and am not the only one with this opinion, that the enterprise on Antwerp would not only have been possible, but indispensable. Antwerp was then without defense. In addition, if Allied troops were sent to it that would have only meant more troops to capture. The distance between this fortress to Brussels obliged the Allies to extend themselves so as to look after the defense of these types of line made their position too weak to hold it, if even the ability to hold it, against our laying bridges over it or fixing our attacks at any point. This region is extremely covered on our side, which is higher at intervals and gives our operation every chance of success. One was, from the beginning of August, in a position to know what to do, be it by the happy success of an attack or by the retreat of the Allies behind the Rupel. One or the other would have put the army in a state to march on Ostend those battalions destined to besiege it. Thus this siege would not have been delayed by the number of operations. The trench would have been opened around 16 August and no later than 25 August. This arrangement supposes that the troops destined for Ostend would not have left the army until after the reduction of the city of Antwerp, whose castle, which was without defense, would have soon succumbed to the same fate.

"I do not know, but I believe that this project found protectors in the council. But they were not sufficiently powerful so as to prevail. The capture of Ostend dazzled the world. It was thought that there was not a moment to lose and that there barely remained many good days to achieve the goal. One did not even want to reflect on the bad condition of this fortress which at least gave them time not to despair. I suppose that one knows perfectly and could not imagine that it could be ignored in a council where the state of the enemy was the base of all reasoning and politics. The siege of a fortress in this state assuredly does not require a month and a half. Events have proven this.

"To show to the end the solidity of the project, I want to suppose for a moment that they occupied our forces long enough to make us unable to undertake the siege of Ostend and that this fortress remained in the power of the Allies, instead of Antwerp, which we had taken from them. I ask what would Ostend be for them in the present situation and what could not have been done by Antwerp, whose surrender would surely produce that of Brussels and all of Flanders reducing the enemy army to Mons, Namur, and Charleroi for all winter quarters? By which port would the English move who were destined to assist the King, the Elector of Hanover? From where would the assistance destined for Ostend come? Is it possible that Holland would have granted them passage, having 100,000 men, so to say in the heart of its states? The event today makes me reason in this manner; but at the time where this showed with certainty the fear that we inspired in Holland, the certainty that the English had no trade agreement with their country, neither for recruits, neither for corps of troops, neither for clothing, nor for any military supplies, should have reduced Holland to refuse them passage for all these things, the open possession of such a rich canton and the extraordinary closing of an army in its winter quarters should have, in my opinion, overcome those who caused the contrary opinion to prevail."

CHAPTER IV
The Prolongation and the End of the 1745 Campaign.

"The departure of the King," wrote de Vault, "appeared to announce that there would be no question of important operations for the rest of the campaign and that one had nothing to do but to occupy themselves with the means of living at the expense of the enemy and of establishing themselves solidly in winter quarters in the region. However, several political and military objectives engaged the King to wish that his army not remain inactive and the Maréchal de Saxe should profit from the rest of the good weather to

Louis XV by Louis-Michel van Loo, 1759

bring a glorious end to the campaign that he had so far executed."[452] If, in effect, the campaign in Flanders appeared to end at the beginning of September, certain events would unroll in another theater that would cause it to continue.

There was no activity in Germany. Conti's retreat to the left bank of the Rhine had closed operations. Otherwise, the denouement of the electoral comedy then playing out in the Frankfurt Diet was no longer doubtful. The election of the spouse of Maria Theresa remained certain and the French would suffer a definitive check in this arena.

However, an unanticipated compensation offered itself in another theater. The Pretender Charles Stuart had landed in Scotland. This event produced a considerable emotion in England. The London Court did not doubt that this new Jacobite effort was supported by Louis XV from the month of September the Pretender became

---

[452] De Vault, p. 66.

so threatening that the British forces remaining in England appeared insufficient to defend the dynasty from the revolution. This circumstance permitted the hope that France would find an occasion to strike King George; because if the Elector of Hanover was safe from French efforts, the English crown appeared to wobble on the King's head. The King of England did not have sufficient forces to immediately smother the efforts of the Pretender. He hoped, with the campaign over, to be able to recall part of the forces that he was maintaining in Flanders. It sufficed, therefore, to continue active operations to make George II's situation most cruel and to support, at the same time, the Jacobite Revolution. Louis XV resolved to do just that.

The "views of His Majesty" and the "political and military interests" to which de Vault referenced found their development in a letter that the Maréchal de Noailles addressed to Maurice de Saxe by the order of the King. After having shown all the importance of the events in England that would subsequently oblige George II to recall certain English troops from Flanders; after having insisted on the emotion of the Dutch, who, distressed by the approach of the French, reinforced the fortresses on their frontiers, Noailles transmitted the Royal desire "to bring to a peak the glorious campaign" which was then wrapping up.

"It is less a question of following," he added, "in that which one can propose, the purely military regulations than political principals. There are some occasions in which it is wise to pass above the ordinary regulations to procure more advantages more considerable than the disadvantages that one might fear." After this most obscure preamble, Noailles proposed nothing less than capturing Brussels and Antwerp and using them as the French winter quarters. It summarized the desiderata of the Court in a formula that was as simple as seducing: There shall be nothing between the French frontier and that of the Dutch. Finally, what remains of French troops in Flanders, be it of the Queen of Hungary, be it of the Allies, shall find itself reduced to enclosing in the three or four strong points what this Princess still retains in this country, or to retire far from this frontier.

Maurice de Saxe responded ten days after these overtures. He responded in terms sufficient to establish his opinion was not to undertake new operations and that he awaited formal orders from the Court to determine his next actions. The Marshal estimated that it could not be a question of taking neither Brussels nor Antwerp. The sieges of these cities did not permit leaving the troops the time necessary to make their repairs before the resumption of the next campaign. They did not assure us, otherwise, more than the possession of which the Marshal regarded as impossible to winter the army in total security.[453]

At this time, the Dutch caused a serious incident that the Minister of Foreign Affairs brought to the attention of Maurice de Saxe, hoping, no doubt to decide the Marshal to undertake the operations for which he showed so little enthusiasm.[454] The Republic appeared resolved to allow nine battalions to move to England for the service of King George. Among these contingents was the former garrison of Tournai. That formation could not, according to the terms of the capitulation, serve against France or her Allies, nor be employed in any military function for 18 months. It was also forbidden for these troops to serve in a garrison. Louis XV prepared to make presentations before the States General of this violation of their engagements.[455] Finally, d'Argenson, Minister of War, took part in his turn in the conversation engaging with Maurice de Saxe and transmitting, on 14 September, the Royal intentions to pursue the campaign, in further specifying them.[456] The Minister asked, in effect, that Maurice prepare an enterprise such as the siege of Ath or another of the same type of his choice. "It is important," he added, "that one not think the campaign in Flanders is finished. The impression that this will make on the King of Prussia, under the circumstances where he finds himself, has given rise to complaints based on political reasons that would take too long for you to infer and about which you are sufficiently informed so that you can make up for them by your own reflection." It appears that it would be very inopportune to evaluate this last argument at the time when, without informing the French, the King of Prussia concluded the Hanover Convention with England. The victory that Frederick had gained at Freiberg had made him very threatening for George II's electorate. Obliged to return to London, George II overcoming his repugnance had signed a ratification, with Frederick II, of the Treaty of Breslau to assure the security of Hanover. At Versailles, they had yet only to fear the signing of such an agreement. The French Ambassador to Berlin sent very little on this subject beyond timid insinuations instead of

---

[453] Maurice de Saxe to Maréchal de Noailles, Alost, 10 September (Corresp. A.F.).

[454] Marquis d'Argenson to Maurice de Saxe, Versailles, 13 September (Corresp. A.F.).

[455] The representations of the Abby de la Ville, the French Ambassador to Holland, remained without effect.

[456] D'Argenson to Maurice de Saxe, Versailles, 15 September, (Corresp. A.F.).

pressing and sure information, which was not sufficient to disillusion such an informed spirit as that of d'Argenson in favor of the Prussian alliance. The response of Maurice to Noailles established, however, that the clairvoyant eyes were already open to the new Prussian defection. D'Argenson remained blind and of him, as of the Prince of Anhalt, Frederick would say, "It is of these men who are narcissists in their opinions and always abundant in their sense of self."[457]

Be that as it may, Maurice responded, on 18 September, to the Minister, that he was going to make all the preparatory dispositions in light of the siege of Ath. He persisted, nonetheless, in disapproving the operations that he prepared only to undertake "by obedience." In sum, the Marshal asked for formal orders. On 21 September, d'Argenson expedited them from Paris. He added to them a little restriction done to dissipate the repugnance of the Marshal. The King related them entirely to his general; but "it is on the success of the core of the enterprise that His Majesty would not like to suffer the least uncertainty in this event. The importance of the capture of Ath did not appear to him sufficient to expose his arms to suffer a check at the end of such a glorious campaign."[458]

"It is very easy for a commander when one has hot feet," Maurice said, in a moment of bad humor, upon the reception of these orders, adding "that it was not necessary to know the land and warfare to order the siege of Ath."[459] Other general officers shared this perception.

Maurice de Saxe attached no tactical importance to the capture of Ath. In the course of the negotiations undertaken to decide this operation, Noailles had stated, in the last argument, that he acted, basically, "to please the Court."[460] This reason appeared insufficient for Maurice. It appeared to him that simple demonstrations would suffice to maintain the English troops in the field in Flanders as it was proposed, without making it necessary to undertake the important operation that was being imposed on him.

It is necessary to consider, in effect, that following the tactics of the time the siege of Ath would oblige the Marshal to displace the army so as to establish a screen between the fortress and the Allies. The

[457] Cf. Broglie, Vol. II, p. 314.
[458] D'Argenson to Maurice de Saxe, Paris, 21 September (Corresp. A.F.).
[459] Godefroy, 21 and 25 September.
[460] Maréchal de Noailles to Maurice de Saxe, Saint-Germain, 17 September (Corresp. A.F.).

French troops would have to camp in a new region that had already been partially exhausted and where they could only live with difficulty. The formal prescription sent by d'Argenson made the necessity to abandon the region around Alost even more urgent. He thought that the Allies would not decide to move away from Brussels, from the Vilvorde Canal, and from Antwerp to attempt a relief of Ath because, he said, "this is the sensible thing."[461] Three days of march were necessary for the Allies to move from Brussels to Ath, while it would only take two for Maurice to move from Alost to Brussels, "where," added the Marshal, "I was placed between this city and the Soignes Forest; the enemy could only come to me except by the defiles that touched, so to say, the gates of Brussels."[462]

"Besides, forcing to abandon to de Clermont-Gallerande (chosen by Maurice to direct the operations against the fortress) the siege of Ath, in this position he can join me by moving along the left of the Dender and then the Allies would find themselves obliged to make a long detour through a very difficult country by leaving the Soignes Forest on their left with very few means for their subsistence." One will soon see by the criticisms of the friend of the Count d'Estrées that the Marshal could not depart, with impunity, from the tactical customs of the time.

Foreseeing, no doubt, the outcome of the negotiations that had begun on 1 September, Maurice had begun some preparations. He had ordered de Brézé to expedite an artillery train from Tournai to Dendermonde by the Escaut, with an eye towards dividing the Allies' suspicions and maintaining their concerns for Antwerp. The 40 cannon and 20 mortars actually destined for the siege of Ath were not to be dispatched from Tournai until the opening of the trench; that is to say, the moment when Maurice would be sufficiently informed on the projects and intentions of the generals of the Pragmatic Army. If these latter persisted in not moving their lines, de Clermont-Gallerande would receive this artillery by Leuze in four successive convoys, gradually according to needs. As for Clermont-Gallerande, he had orders to move to Lessines with his detachment on the 22nd, and to invest Ath immediately. As Clermont-Gallerande had very little infantry, Maurice quickly sent him a reinforcement of the necessary battalions.[463] D'Estrées finally moved to take up a post at Enghien

---

[461]Maurice de Saxe, Alost, 23 September (Corresp. A.F.).

[462] *Ibid.*.

[463] Maurice de Saxe to d'Argenson, Alost, 21 September. (*Ibid.*).

with an important corps of dragoons and cavalry to observe the Allies' movements. Such were the general dispositions taken by the Marshal. He estimated that the active phase of the siege of Ath would not begin before 1 October. "You have the time from now to then to send me new contrary orders," he said to d'Argenson, informing him of his project; "but if you do not send me other orders, I shall go forward with the plan contained in this letter."

It was unfortunate that the Court did not ratify these prudent, simple, and judicious propositions. On 27 September, d'Argenson responded that the King "extremely approved the plan of conduct laid out for the decision on the siege of Ath."[464] The negotiations begun between the Court and the Marshal had lasted a month and during this time other events had drawn the attention of the French to their frontiers.

The free troops and partisans that we have seen coming from the Allied lines at the end of August had not delayed in imposing numerous exactions. With the assistance of the garrisons of Mons, Namur, and Charleroi, these Allied pillagers spread terror throughout the region of Givet, Maubeuge, and Philippeville. During the night of 3-4 September, 600 infantry and 300 cavalry arrived at Beaumont. The following day they pillaged Barbençon and Bossu and carried off several mayors. This blow done, they retired with their booty towards Mons and Charleroi. The population had raised the alarm at Givet and Mariembourg. The governors of these places did not have sufficient troops, especially cavalry. As a result they could not even harass the Allies' retreat.[465] At Maubeuge, Phelippes was in a better state to resist. He had been informed of the approach of about 1,000 pillagers. He immediately sent to engage them the partisan Goderneaux and his free company of dragoons, two companies of grenadiers, a battalion of militia, and two parties of the Provisy and Marsanne Free Companies. As at the moment the King was returning to Paris, Phelippes had

[464] D'Argenson to Maurice de Saxe, Paris, 27 September, (Corresp. A.F.).

[465] De Fiennes (lieutenant of the King) to d'Argenson, Givet, 5 September. – D'Ingouville (lieutenant of the King) to d'Argenson, Philippeville, 6 September. – Phelippes to d'Argenson, Maubeuge, 4-5 September (Corresp. A.F.). – According to the Army return of 13 August, already cited, the garrison of Maubeuge contained no more than a battalion of militia, 120 fusiliers detached from the Marsanne Free Company and the Provisy Free Company, with 150 dragoons from the Goderneaux Free Company. The garrison of Givet contained 4 squadrons of the Talleyrand Regiment and 1 from the Grammont Regiment distributed between the fortress and its vicinity with 150 fusiliers of the La Gaigneur Free Dragoon Company.

informed the Minister of these Allied incursions, adding that it was almost impossible for him to fight because of his lack of troops. The King sent this letter to Maurice de Saxe. Louis XV thought that his march on Paris would not be troubled by this Allied operation. However, he wrote to the Marshal "to not tolerate that the enemy put our frontier to the contribution without obstacles, while we are masters of the greater part and the best part of the Queen of Hungary's lands."[466] As a result, Maurice resolved to reinforce the garrisons on the frontier. Nothing opposed this project since the army, at least one thought so at the moment, was about to go into winter quarters. On 10 September, two infantry regiments (Bouzols and Fleury) left Alost for the camp of Clermont-Gallerande at Enghien. This latter was to attach to them the Mestre-de-Camp Cavalry Regiment and the Septimanie Dragoon Regiment, reduced to their own effectives. The ensemble of these troops were to be spread along the frontier such that Fleur and Septimanie occupied Maubeuge, two battalions of Bouzols and two squadrons of Mestre-de-camp occupied Beaumont (a small fortress of the Queen's on which one supported almost all the Allied expeditions), and the rest in Philippeville.[467] At the same time, three maréchaux de camp from the army, Relingue, la Motte, and Gravel, went to, respectively, Beaumont, Philippeville, and Givet, in the capacity of governors under the authority of Lieutenant General Phelippes, commandant of Maubeuge. These officers were to take, according to Maurice's orders, all the measures necessary to oppose the raids of the Allied and to draw "contributions" from the region of Chimay as a reprisal.[468]

On 11 September, the Marshal dispatched a new column for Enghien. It consisted of two regiments of infantry, one of dragoons, and 10 Swedish guns destined to fill the holes produced in the detachment from Clermont-Gallerande by the formation of the reinforcements destined for the frontier.[469]

---

[466]D'Argenson to Maurice de Saxe, Lille, 5 September (Corresp. A.F.). – Maréchal de Saxe estimating that he lacked the fresh troops in the army sent the order to the commander of Maubeuge to send the Marasanne and Provisy Companies to his camp. Phelippes responded that he was going to send them as soon as they returned from a mission, whose details he provided. (Phelippes to Maurice de Saxe, Maubeuge, 4,5, & 6 September). – The Marshal received, at the same time, a letter from d'Argenson cited above and those of Phelippes, which gave him the necessary explanations.

[467] Crémilles to d'Argenson, Camp at Alost, 9 September (Corresp. A.F.).

[468] *Ibid.*

[469] Crémilles to d'Argenson, Camp at Alost, 11 September (Corresp. A.F.).

Henry Pichat

To escort these last troops to the points they were to occupy, Beausobre and his hussars and the mounted Grassins left Lessines, on 12 September. The following day, finally, Clermont-Gallerande and his detachment marched out in their turn to move on the Austrian Hainaut and the Comté of Namur.[470]

"The Marshal had given, as the sole objective of Clermont-Gallerande's movement to support the corps that de Beausobre led and which was only a day ahead and that at the same time was to spread terror in the land as well as arrest some bailiffs," wrote Crémilles to d'Argenson.[471] In summary, Beausobre marched with his light horse and the reinforcements destined for the fortresses, the latter in the advance guard. Clermont-Gallerande and his detachment were to follow and support Beausobre until the latter's column crossed the Sambre. To fulfill his mission, Clermont-Gallerande pursued his path across the region of Méhaigne and the Meuse. He returned, finally, over the territory between the Sambre and Meuse, having his troops live at the expense of the territory he crossed. Following the ordinary method, Crancé (commissioner performing the functions of the intendant of this corps) sent the necessary orders. Beausobre used his light cavalry to execute the requisitions.

In sum, in view of the retreat of the army to the left bank of the Dender, and also the mission given to Clermont-Gallerande, the reinforcements sent from Alost by the Marshal were not wasted. The prescriptions of Maurice assured Clermont-Gallerande a liberty of maneuver which no one had enjoyed during the campaign. It was, perhaps, a bit late for this latter to completely fulfill the mission he had been given. The negotiations engaged between the Marshal and the Court would soon sensibly reduce the length of the foreseen operations.

Be that as it may, Clermont-Gallerande left Enghien, on 13 September, sending de Maugiron to prepare for marches and encampments.[472] The day before, Beausobre arrived in Feluy and Arquesnes with the reinforcements destined for those fortresses, the hussars and the Grassins. The latter light troops had come from Lessines in a single push.

[470] *Ibid.*
[471] *Ibid.*
[472] Maugiron sent to d'Argenson, dating from 27 September and from the camp below Ath (Corresp. A.F.), the journal of the marches of this detachment for the period of 13-26 September. We have taken from this document the details of the marches that follows.

The mission given to Clermont-Gallerande imposed on him a need to march rapidly. If the regiments sent to Maubeuge, Beaumont, and Philippeville could not lighten themselves of their baggage, nothing would prevent the return to Lessines, where Clermont-Gallerande was to send the heavy baggage of the other units. It appears that Clermont-Gallerande had the weakness to go because of the protests of his officers. The columns loaded up all their baggage and the convoy of baggage echeloned interminably along the roads.[473]

The first post in the march of Clermont-Gallerande led to Feluy and Arquesnes.[474] The attention of the Allies was awakened. On 12 September, the Allies learned of the reinforcement of the French troops camped at Enghien. They suspected that it was the movement on Namur that Beausobre would then invest.[475] Waldeck sent his adjudant, Cornabé, on a reconnaissance with two free companies, 100 dragoons, and a company of grenadiers. These troops left the Allied camp, on 13 September, and after having sent some parties to observe the Braine and Soignes Forests, Cornabé camped at Braine-la-Leud.[476] The baggage of Clermont-Gallerande's officers had been placed in the tail of the central column and covered by the two wings marching in line of battle by companies. The Allies did not dare to strike it.[477]

To his great surprise, upon arriving at Feluy, Clermont-Gallerande found Beausobre and his advance guard there, even though they should have departed earlier. Beausobre claimed that the day before the move from Lessines he had been obliged to leave his light troops strongly occupying Soignes as a precaution; their march was singularly long and that after being 23 hours in the saddle they had not left until the morning of the 13th. Would this first setback oblige Clermont-Gallerande to rest in Felury in order to permit Beausobre to resume his advance? The disadvantage of this loss of time almost immediately disappeared, as we know, because Clermont-Gallerande was advised

[473] R.A.

[474] The troops marched in three columns. That of the center contained the infantry, the artillery, and the baggage. It crossed the Little Senne at Rebecque and crossed Braine-le-Comte and the Escaussines. That of the right, formed of the cavalry and a regiment of dragoons crossed Hoves and reached Fleuy by going cross-country. That of the left, formed like that of the right, left Enghien by the Warelles Castle and thus crossed the fields to move on Feluy. The camp was established in the bend of the Senne where the deep quarries touched Feluy and they then occupied Arquesnes.

[475] Waldeck, 12 September.

[476] Waldeck, 13-14 September.

[477] Maugiron to d'Argenson, camp below Ath, 27 September (Corresp. A.F.).

by the Marshal that the duration of his mission had been reduced, but nonetheless it remained urgent for him to push forward the reinforcements destined for the various fortresses.[478] Clermont-Gallerande adopted a new disposition. He decided that everyone would leave the following day, the 14th. He would stop at Celles, but Beausobre and his troops would camp at Ligny. This latter would link himself to his chief by means of small posts. In this position, Beausobre would continue to deceive the Allies by capturing several mayors as far as the gates of Namur while Clermont-Gallerande marched towards Saint-Gérard, threatening Charleroi. At this moment, the reinforcements destined to the various fortresses crossed the Sambre protected on one side by Beausobre and on the other by Clermont-Gallerande.[479] A second mishap would also modify this new project.

On 14 September, the march resumed. Clermont-Gallerande reached Celles without encountering any obstacles, following in the steps of Cornabé, who camped at Genappe. During the night of the 14th/15th Beausobre sent a courier from Ligny at 11:00 p.m. He carried word that the Allies were marching on the French advance guard. As a result, Beausobre, fearing he would be attacked in the early morning, crossed the Sambre and camped three miles from Ligny.[480] As a result, Clermont-Gallerande found a new delay imposed on him. He could not think of moving on Saint-Gérard before knowing the exact situation, as the information from Beausobre was insufficient. Clermont-Gallerande rested, the 15th, at Celles and sent a reconnaissance towards the Soignes Forest, the only dangerous side. The reconnaissance only saw Cornabé marching from Genappe on Villers. Thinking then that Beausobre could cross the Sambre and, as a result, deliver the reinforcements under cover from all attack, Clermont-Gallerande withdrew to Binch, on the 16th. He considered his mission as finished.[481] This movement did not escape Cornabé, who reported to

[478]R.A.

[479] Clermont-Gallerande to d'Argenson, Camp at Chièves, 16 September (Corresp. A.F.).

[480]*Ibid.* – To move from Feluy to Celles, Clermont-Gallerande formed his troops in three columns, formed as the day before. The center moved about 3 miles along the road from Felury to Celles. The right wing moved directly cross-country on Celles. The left did the same, leaving the Renisart Wood on its right to cut across to the center. The troops occupied a camp covered by the Piéton and supported on one side by the bend in the stream at the point known as Blanc-Cheval and on the other on the Celles Bridge.

[481] Waldeck, 16-17 September. – What did Beausobre do? Did he arrive at Walcourt on the 16th, or on the 17th? If he decamped from Ligny as rapidly as his letter sug-

von Waldeck, adding that that same day Beausobre had crossed the Sambre at Château and camped at Walcourt, on the right bank of the river. Cornabé reported, in addition, on the 17th, the presence of the French light cavalry in this region.

On the 16th, Clermont-Gallerande arrived in Binch and waited for Beausobre, who joined him on the 19th. Clermont-Gallerande received, on the 18th, his resupply of bread coming from Maubeuge.[482] The Marshal, having directed him to be in Lessines on the 22nd, nothing prevented Clermont-Gallerande from resuming his march on the 20th.[483] However, he did not leave Binche until the 23rd. De Machault, Intendant of Hainaut, had informed Clermont-Gallerande that the Allies had stopped a courier; but he had forgotten to say that he was acting as a mail postilion. Clermont-Gallerande thought that the Allies had captured a messenger from the Marshal bringing new orders and he waited at Binch for confirmation of his initial instructions.[484] Greeting with the same alacrity the absurd rumors and useless opinions, which came from Beausobre and Machault, Clermont-Gallerande lost precious time.

To make up for his delay, he resolved to execute in two marches the distance that would normally take three days over such broken ground that was little suitable for long marches. On the 23rd his troops covered six difficult miles.[485] On the 24th they completely exhausted themselves in yet another longer march.

The detachment reached Lessines in the greatest disorder.[486] A good reconnaissance of the roads might otherwise have alleviated, in part, the inconvenient results of the abnormal elongation of the march stages. Clermont-Gallerande took no measure, however, in this

gests, why did he not conform to Clermont-Gallerande's orders which directed him, under such circumstances, to withdraw? Did Beausobre think he had completed his mission? Why then did he not inform his general of this rational decision which he appears to have taken under these circumstances? Always well inspired, R.A. insinuates that at this occasion Beausobre was of service to both Clermont-Gallerande as well as the Marshal, because his maneuver shortened the "march" of the first and served, at the same time, the plans of the second. R.A. sees, otherwise, a confirmation of this bold hypothesis in the impunity reserved to two "quarrels", according to his own expression, of the commander of the advance guard.

[482]R.A.—Clermont-Gallerande to Maurice de Saxe, Camp at Chièvres, 16 September (Corresp. A.F.).

[483]Maurice de Saxe to d'Argenson, Alost camp, 21 September. (*Ibid.*).

[484]Maurice de Saxe to d'Argenson, Alost camp, 23 September. (*Ibid.*).

[485] Maugiron to d'Argenson, camp before Ath, 27 September. (*Ibid.*).

[486] R.A.

regard. It appears that the march of his troops was hastily prepared, because of the precipitous departure from Binch after the renewal of the Marshal's orders, be it because of the weather, the means or the experience which was lacking in Maugiron. Be it as it may, the detachment camped, on the 23rd, at Cauchies-Notre-Dame and, on the 24th, at Lessines.

Recalled by Waldeck, Cornabé returned to the Allied lines. "We were," wrote the Prince "always well informed as we have seen several times, that Clermont-Gallerande's corps had for its objective capturing the garrison of Ostend when it left Mons to rejoin the army.[487] Beausobre's corps had for its objective demanding new contributions from the region, happily, he thought that the detachment commanded by my adjutant to observe him was larger than it was. That obliged him to cross the Sambre and retire to Binch!" In sum, Cornabé had seen nothing and for the instant, Waldeck was reassured, five days later he would be afflicted by cruel concerns when he learned that Ath had been invested by the French.[488]

Thus, according to the dispositions made by Maurice de Saxe, Clermont-Gallerande and Wurmbrant, the Austrian Governor of Ath, had fought between two observing armies a singular combat. The fight was not epic. Wurmbrant deployed in his defense the energy and activity which one could expect from an old man of 80 years. As for de Clermont-Gallerande, he added nothing to the renown he had gained from his earlier services.

On 25 September, Clermont-Gallerande received his last instructions from Maréchal de Saxe at Alost.[489] Maurice decided that the investment would begin on the 26th. The French had every advantage to gain from taking their time as they would learn that the fortress contained only 1,500 defenders and that its artillery had not yet been mounted on carriages. The approach of bad weather encouraged the French to hasten their activity. One had no reason to expect that

---

[487] The Allies desired to recall to their lines the English troops directed on Mons after the capitulation of Ostend, because the London Court was going to recall to England the British contingents. According to Waldeck, on 24 September, 5 English battalions left for Antwerp, followed the next day by 5 others and by General Ligonier. All these troops moved on Wilhelmstadt to embark for England. In compensation, a first reinforcement of Hessians arrived at Malines on the 25th. On the 29th, the Allies awaited a second division of these troops.

[488] "One doubts that the enemy wishes to [attack] Ath." (Waldeck, 24 September). – "One has confirmed the investment of Ath." (Waldeck, 25 September).

[489] Crémilles to d'Argenson, camp at Alost, 24-25 September (Corresp. A.F.).

the Allies would send reinforcements to Ath.[490] It was, subsequently, agreed that d'Estrées would occupy Enghien, on the 27th, with the 23 squadrons he commanded, to which Clermont-Gallerande added the 5 squadrons of dragoons drawn from his numerous cavalry. D'Estrées was given the mission of observing the Allied movements on the side of Brussels.[491]

On the 26th, Clermont-Gallerande left Lessines with his 8 battalions and 33 squadrons to invest Ath. The following day the Marshal sent him, from the Alost camp, 23 battalions.[492] This reinforcement camped at Grammont, on the 27th, and rejoined the siege corps, on the 28th. With an eye towards filling the holes created in the army by the departure of this infantry, Du Chayla received orders to assemble, under Dendermonde, 9 battalions and 18 squadrons, and to hold them ready to join the Marshal on the first order.[493] The artillery expedited from Ghent was to find itself in Tournai, on 1 October. The engineers were gathered on this date before Ath. Finally, the Commissioner Crancé assigned by Séchelles to the siege corps, received detailed instructions regulating the details of his service.[494]

On this day, Clermont-Gallerande arrived before the fortress with his detachment and "remained mounted until 9:00 p.m. to emplace it." These troops bivouacked while they awaited the arrival of the reinforcement of 23 battalions. The repair of the road from Leuze and the various roads leading to Tournai, the center of resupply for the siege corps, began on this day as well.[495] The infantry sent by the Marshal camped, on the 27th, between Lessines and Grammont along the road linking these two cities. It resumed its march on the 28th. At Rebaix it found local guides to lead each of the different corps over the shortest road to the emplacements that had been reserved for them in the line of investment.[496] De Clermont-Gallerande had done nothing new. He had adopted the same disposition used by the Allies in 1706. Considering that the broad expanse of ground between Leuze Stream and the Lower Dender was covered by the army and the posts of the

---

[490]Clermont-Gallerande to Maurice de Saxe, camp before Ath, 26 September, 7:00 p.m. (*Ibid.*).

[491] Crémilles to d'Argenson, camp at Ath, 24-25 September. (*Ibid.*).

[492] *Ibid.*.

[493]*Ibid.*.

[494] Crancé to d'Argenson, Arbre, 27 September (Corresp. A.F.).

[495] Clermont-Gallerande to d'Argenson, camp before Ath, 28 September (*Ibid.*).

[496]Ibid. – La Tour (chief of staff of the siege corps) to d'Argenson, camp before Ath, 28 September (Corresp. A.F.).

Leuze Grassins was watching Mons, Clermont-Gallerande had placed only six battalions and four squadrons in this region to hold the principal roads. By an unusual innovation, however, the cavalry occupied the Renard Wood and the infantry occupied the Arbre plain.[497]

On the 29th, it was ordered to prepare 500 fascines, 50 gabions, 50 wicker screens, with sap faggots per battalion and to transport these to the depots organized at Masse and Irchonwels. The following day there was still nothing in the magazines and it was seen with this slowness of work one could not hope to open a trench before 2 or 3 September. During this time, the Allies showed their artillery and began to burn the houses in the suburbs to clear the Mons Gate. The dispatch of some grenadiers sufficed to interrupt this effort.

It was decided to attack Ath by two fronts simultaneously, one from the left bank of the Dender at the Tournai Gate and the other over the ground between the Dender and the Masse stream. On this last point, where Clermont-Gallerande renewed the effort of 1706 against Ath, the true attack would come.[498] The other attack was to only be a feint pursued with vigor in order to facilitate the beginning of work on the main attack.

On 1 October, 1,200 workers and their support, formed of 3 companies of grenadiers and a battalion of infantry, were assembled at 7:00 p.m., at the depot at Villers-Notre-Dame and began to remove dirt on the side of the false attack. They constructed a trench running from the Dender to the Tournai Highway and linked it to the depot with a communication defile behind a screen of trees. At daybreak, a battalion occupied the trench.[499] Two hundred masters[500] assembled in under cover near the Tournai Gate to defend against a possible sortie by the garrison. A misunderstanding resulted in them not occupying their post until daybreak. This was not a problem, however, as the garrison had focused all its attention on the side of the false attack. The garrison had, nonetheless, developed some suspicions. Towards 11:00 p.m., they threw some pots au feu on the side of the false attack, where the where the workers had already entrenched themselves.[501] The French peacefully constructed a battery on this side with a view to

---

[497] R.A. – La Tour, in the letter cited above, sent to d'Argenson in detail of the emplacement of the troops of circumvallation, which justifies R.A.'s assertion.

[498] Clermont-Gallerande to Maurice de Saxe, camp before Ath, 1 October (Corresp. A.F.).

[499] La Tour to d'Argenson, camp before Ath, 2 October (Corresp. A.F.).

[500] Translator: A "master" is a heavy cavalryman.

[501] *Ibid.*

enfilading the defensive front.

During the evening of 2 October, 3,000 workers protected by two battalions of infantry and four companies of grenadiers dug in at the point of the true attack.[502] The work began at 8:00 p.m., and at 11:00 p.m. everyone was dug-in. According to custom, running from the two depots, the engineers constructed two cuts which when united formed the first parallel. Communication trenches of 150-200 toises linked Irchonwels to the camp. This parallel ran about 600 toises and was about 110 toises from the fortress. The great sound carrying over the ground from the false attack drew the fire of the Allies to that side so well that the workers were not disturbed.

On the 4th, four batteries were completed, the equipment had arrived from Tournai, and the artillery of the false attack opened fire. It was necessary to stop firing for two hours in order to thicken the parapets which offered insufficient protection. After this the fire resumed causing much damage to the fortress. On the side of the true attack the cannon only entered action during the night of the 6th; but on the 7th, 21 cannon and 11 mortars began firing.

The works pushed forward. A second parallel was completed. The flying saps left from the left soon approaching the tenaillon of the covered road and that of the right descended the forward ditch of a flèche protecting the demi-lune that covered the road. The French constructed a bridge of fascines destined to cross the ditch.

During this time, the Allies seemed to act in their lines behind the canal and certain events, which will soon be addressed, caused the French to fear that the Allies would not attempt the battle that the Marshal had refused to "respond to" in his last latter to d'Argenson. Maurice resolved to rush the outcome and on his orders Clermont-Gallerande wrote to the Governor of Ath to engage him to capitulate before being reduced to the last extremity. Clermont-Gallerande's request had no success.

During the duration of the negotiations, one saw with surprise that the bastions on the front of the defense were counterguarded. The garrison did not appear disposed to make use of all its means of defense, because during the night of 7/8 October the flèche covering the left bastion, abandoned by Wurmbrandt's soldiers, fell into French hands without a fight. [503] During the morning of the 8th, the French projectiles began to burn in the city. The fire burned the house of the

[502] La Tour to d'Argenson, camp before Ath, 2 and 3 October (Corresp. A.F.).
[503] La Tour to d'Argenson, camp before Ath, 8 October (Corresp. A.F.).

governor, next to the arsenal. This produced a certain panic and one could hear the garrison sound the general [alarm] and the tocsin to call for assistance against the fire. It appeared that this instance eliminated Wurmbrandt's last scruples, because he raised the white flag at 1:00 p.m.

That same day the capitulation was signed.[504] Clermont-Gallerande granted the honors of war to the entire garrison. Maréchal de Saxe began to run out of forage in his camp at Alost. Fearing to prolong Wurmbandt's resistance by being demanding, Maurice ratified the articles signed by his lieutenant. [505] Otherwise, the rapidity of this capitulation singularly served the Allies. Louis XV had wished that the Dutch of the garrison of Ath be disarmed and interned in France as a reprisal for the sending of the garrison of Tournai to England. [506] The letter from the Minister, that informed the Marshal of the Royal intentions, did not reach him until 14 October, the date on which the garrison of Ath entered Brussels.[507] De Clermont-Gallerande, we have seen, was not unaware of the projects of Louis XV with regards to the Dutch. However, not having received a formal order, he did nothing to satisfy the Royal intentions.[508] The friend of the Count d'Estrées appreciated, in his habitual manner, the facility with which Wurmbrant had calmed the scruples of his conscience. "He had no other pretext," he said, "to cover his cowardice than the burning of the arsenal; but to suppose that the munitions had completely burned up, he had only to draw us out for eight days and the fortifications of his fortress were not capable of preventing him from capitulating."[509]

Thus ended the last siege of the campaign. Of Wurmbrant, like the other Austrian governors that the French had attacked, a contemporary could justly write: "When one has an affair with equal defenders, one can attempt anything and be certain of conquering every-

---

[504]*Ibid.*.

[505] *Ibid.*.

[506] *Ibid.*.

[507] D'Argenson to Maurice de Saxe, Fontainebleau, 9 October (*Ibid.*).

[508] The prescriptions of the Minister reached Maurice at the Baerleghem camp, on 14 October. In his accusation of the reception, the Marshal wrote: "You shall judge by the example of the disorder of the posts. Your letter of the 9th did not reach me until the 14th. The correspondence to Flanders is as bad as always. A letter form Alost to Audenarde took five days, one for Ostend took four, and thus for the rest."

[509] D'Argenson to Maurice de Saxe, Fontainebleau, 9 October (Corresp. A.F.), see p. 210, note.

thing."[510]

After having related, without interruption, all the events of the siege of Ath so as to feed the clarity of the account, it is now necessary to return to the other events of Clermont-Gallerande's account that occurred during the first days of October.

We have shown that at the end of September, the London Court had assembled in Willelmstadt a certain number of troops destined to return to England to reduce the Jacobites. It resolved to join to these troops the five English battalions that had formed the old garrison of Ostend, which had been sent to Mons after the capitulation. From 11 September these British troops had been in Mons, but the Duke of Cumberland had not dared to send them out. He feared, as we have said, that the "promenade" of Clermont-Gallerande had no other objective than capturing these five battalions if they imprudently marched along the highway to Brussels, the shortest and most direct road to Antwerp. The presence of an important siege corps below Ath and also that of d'Estrées with 28 squadrons at Enghien from 27 September complicated the situation even more. If, after the capture of Ath, the French moved to besiege Mons, as the rumor ran, the return of the English battalions could postpone *ad calendas* by admitting that they were not made prisoners after the fall of Mons.

From the beginning of the siege of Ath, the Allies resolved to profit, without delay, from the opportunity to bring back the English battalions and to replace them in Mons with four Austrian battalions. On the condition of secretly preparing this expedition and of executing it rapidly, the Allies could hope to execute it while running few risks. They foresaw, in effect, that the siege operations would suffice to occupy Clermont-Gallerande and that the Marshal was too far away to intervene in a timely manner. In truth, d'Estrées established at Enghien, blocked the road with his squadrons, but the Allies saw the presence of this force the only obstacle that could obstruct the projected expedition. As a result, it sufficed that d'Estrés move away from the road for the Allies to realize their project. They made the necessary dispositions to produce this result as soon as possible. Their plan required that it be executed immediately, as after the capture of Ath it would become impractical. By a singular turn of events, the departure of this English garrison from Ostend was going to become for the French, as it had been for the Allies, a cause of great and chimerical alarms. Because of the lack of calmness of those who should not have

[510]R.A.

allowed themselves to be fooled, the Allies succeeded in deceiving the French. Their true intentions escaped the French and it was thought that they sought to break the blockade of Ath or to bring the French to battle where the Marshal would not dare "to respond" to d'Argenson. As one will see, Maurice was not fooled. He refused to take the measures that the general concern demanded. Before this resistance a true panic arose. The agitation was short lived, it is true; but when the events had dissipated, those who had raised it did not pardon the Marshal for not having shared it. They would have, however, avoided it by deploying some of militaries qualities that their friends so liberally displayed to them.

On 28 September, the Allies knew of the arrival of d'Estrées at Enghien. Without delay they sent a reconnaissance in the direction of Braine-la-Leud.[511] The hussars and two companies that formed it began by the capture of d'Estrées' baggage, as it came from Grammont. That he had not made use of the council formerly lavished on Clermont-Gallerande in similar circumstances by his anonymous friend. Moreover, the Allies thus rendered a singular service to d'Estrées. His baggage, without doubt, surely would have embarrassed him a few days later. Lastly, on the 30th, Allied hussars pushed elements to Hall and Tubise, making contact with some of d'Estrées detachments.

On 1 October, a conference was held with Königsegg, "the outcome of which," wrote Waldeck, "was to form a detachment of 3,000 horse, 10 companies of Dutch grenadiers, and 60 hussars under the command of General Howley, who asked me for Cornabé, my adjudant, to lead the advance guard. This detachment departed at 10:00 p.m., and the following morning it arrived at Hal. It had as its goal to attack d'Estrées' corps, or to oblige him to retire from Enghien. After it arrived, Cornabé sent a reconnaissance to study the situation of the ground between Hal and Enghien and, on the relation of which informed him he did not have a sufficient infantry for such broken ground, he sent a request for some more companies of grenadiers, the Scottish Highlanders, and four cannon."[512] This clearly demonstrated the resolution of the Allies.

On 3 October, Howley received two Dutch free companies and 3 English companies; but the reinforcements he had requested did not come until the 4th. The Duke of Cumberland arrived with these reinforcements. The Duke gave Howley, according to Waldeck "secretly

---

[511]Waldeck, 28 September.

[512] *Ibid.,* 1 October.

the order to flush out the corps that was at Enghien and to then move to Mons to favor the exit of the garrison of Ostend which he wished to join the army and send to Mons, in its place, the four Austrian battalions that were with the army."[513] On the 5th, in effect, they arrived at Hal. The day before the English of the garrison of Ostend came out of Mons.[514] Nothing could be more simple, one sees, and there is no reason to doubt the sincerity of Waldeck's account. The Allied maneuver, however, would receive a completely false interpretation from the beginning.

For the sudden and unexpected acts that they committed, the movements of the Allies did not escape the French spies who saw the true objective. The precautions taken by the Duke of Cumberland to only inform Howley confidentially of what the Allies expected of him was most fortunate, at least to judge by the rumors that arose. Immediately advised of the Allied maneuver, though d'Estées still did not know their true objective, he did inform the Maréchal de Saxe of it on 2 October.[515] D'Estrées was indecisive. He appeared to have resolved already to retreat and did not appear concerned enough to verify the important information that he had received. He requested instructions from Maréchal de Saxe. Maurice sent them by return courier. However, the "Clermont-Gallerande's vedette" was to leave the terrain without awaiting orders that had been solicited, as if the events imposed the execution of an immediate action that was so little compatible with the mission that has been confided to d'Estrées.

However, in his answer Maurice indicated to d'Estrées two logical and easy solutions to be employed. "I am persuaded," he wrote, "that your good countenance will cause them to retire to assure the arrival of the Ostend garrison to their army. If before this junction you can, in concert with de Clermont-Gallerande march on this corps and attack separately, this shall be the most wonderful thing in the world. De Clermont-Gallerande shall give you the infantry that you judge you need. If you lack the time and the circumstances are not favorable to this enterprise, one can only let them pass."[516] D'Estrées

---

[513]*Ibid.,* 3 and 4 October.
[514] *Ibid.,* 4 and 5 October.
[515] "At the time our spy arrived from Brussels with the news that 6,000 men had left to attack us and that they werer at Hal."
[516] "Though you had not told him to come, he believed that he should. If he had not been drunk, my determination would have been easily taken; but this small inconvenience held me in suspense while waiting for the return of another spy that did not return. But a trustworthy man arrived from Brussels, who told me that 6,000 men

Henry Pichat

had come to a conclusion, on the 3rd. He would hold it unshakably until the end.[517] At 2:00 a.m., of that day he sent a patrol on Rebecque, Petit-Enghien, and Saint-Renelde. This reconnaissance revealed nothing. At dawn d'Estrées sent out a second reconnaissance containing 40 Grassins, 30 uhlans, 100 horse, and 100 dragoons in the direction of Saint-Renelde. At the same time on his side, Cornabé came with 150 horse, two free companies, and 4 companies of grenadiers to assure that d'Estrées had received the infantry that was rumored in Hal.[518] The French, not particularly vigilant, marched in a column, if one judges by the subsequent events.

When the head of the French column entered the woods near Petit-Enghien, it was fired upon. It then panicked. Without engaging the Allies, the uhlans fell back and broke the dragoons who, in their turn, threw disorder into the cavalry. Only the Grassins, who were not so easily caught, attempted in vain to remain firm. Abandoned, they were obliged to retire by the Allies as far as Enghien. This action and the news that d'Estrées came to receive would determine him to withdraw. He informed the Marshal of this withdrawal. This letter contained the instructions and the advice from Maurice that d'Estrées had solicited, but which he judged pointless to await.

At daybreak on the 4th, d'Estrées retreated. He could hardly have better served the Allies' plans. The French had lost 40 men, as many dead as wounded in the reconnaissance of the previous day. One saw that "the good countenance" of d'Estrées was not very strong. The Marshal quickly received the word of this retreat. The role of

---

had left for Hal and that they would arrive this morning and that they came to withdraw the garrison from Ostend. If it has this objective or for another, I will sound the general at 2:00 a.m. I will return my crews to Ghislenghien. I will put myself in battle formation at daybreak, with my detachments forward to warn me. I believe my retreat is secure and I am at peace. I will inform de Clermont-Gallerande. If they remain at Hal for a long time, what should I do? The union with the garrison of Mons completed, this corps will contain 10,000 men. There will be disadvantages in leaving it too near. If they go away, I will await another alert and your orders in preference to all others.

[517]Maurice de Saxe to d'Estrées, camp at Alost, 3 October (Corresp. A.F.). – This document beings thus: "I have received the letter that you have done me the honor of writing yesterday evening. I have had the same information of the march of an enemy corps on Hal." – Maurice ends his letter as follows: "It is above all necessary that such a corps is under your orders and that of de Clermont-Gallerande remain extremely alert and sure of being warned of the enemy's movements. On this occasion, send your news as frequently as you can."

[518] Waldeck, 3-4 October.

"Clermont-Gallerande's vedette" was ended; d'Estrées would perform another with such activity that would have been better employed on the 2nd, 3rd, and 4th of October. D'Estrées set to work on his task at Ghislenghein, where "he had come to visit with de Clermont and he had shared with him the word that he had recently received that the enemy had brought into their camp at Hal 16 cannon; that he could not rest on the security with which the siege corps of Ath appeared to need. Having convinced de Clermont-Gallerande that it was not prudent to remain thus deprived, when he could find at any moment 25,000 men under him, de Clermont-Gallerande by having little advantage and one more siege to execute, he needed to visit the ground so as to put it in a state of defense. It was Tuesday, 5 October, when d'Armentières and Saint-Perm were charged with the same care on different fronts."[519]

Besides, at this hour, d'Estrées came to share with Clermont-Gallerande the feared news borne in the letters addressed to the Marshal, on 3 and 4 October, since at that date d'Estrées did not believe that the Allies would dare to take the post at Enghien. No event, however, had justified d'Estrées to modify, so radically, his first impression.

Until 2 October, however, Clermont-Gallerande had lived with the greatest tranquility. He appeared completely disposed to remain calm in the middle of which the siege of Ath proceeded.[520] D'Estrées had attempted in vain to disturb this peace on 1 October.[521] It appears that two days later the insistences of d'Estrées began to produce

---

[519]R.A. – The chronicler added: "I followed the Count d'Estrées and had all the leisure to reconnoiter his talents and the accuracy of his military eye which I wished to detail on this occasion so as to not forget it." – R.A. furnished abundant details on the defensive preparations before Clermont-Gallerande.

[520]"We shall have the liberty to execute our siege peacefully. Von Wurmbrandt has made a demonstration in his defense; we will see if it will continue. The best news that can come to me at this time, is the re-establishment of your health, which causes me my only concern." [Clermont-Gallerande to Maurice de Saxe, camp before Ath, 1 October (Corresp. A.F.)].

[521]"I have received, today, a letter from d'Estrées which informs me that there have departed from Mons 3,000 men where you know that there are 4,000 English from the Ostend garrison and 9 battalions which form its garrison. I can give no faith to this news, because I have people in Mons and parties outside it which have sent me nothing. In case that it proves true, as I absolutely wish to prevent them from joining their army or falling on the convoy which must come to me tomorrow from Tournai, in case that I should be warned in times as I hope, I will move with a corps of troops on the Casteau Moor, if they take the road from Soignes, or on Blaton in case they march on Leuze." (*Ibid.*).

some effect on Clermont-Gallerande's spirits. Yet, the later appeared to still persevere in his intentions so boldly manifested long ago to the Marshal to capture the Ostend garrison, if it had ventured outside the walls of Mons. The assurance of these excellent dispositions had, no doubt, encouraged Maurice to recommend the offensive operations mentioned summarily in his letter, of the 3rd, to d'Estrées.

On the 4th, it became evident that the alarm had expanded to the siege corps and that it had singularly grown. "I do not doubt," said La Tour, "that de Clermont-Gallerande did not take sufficient precautions."[522] The Commissioner Crancé represented the general impression in writing: "D'Estrées has withdrawn to Ghislenghein. If the enemies march against him he will retire on us. I believe that de Clermont-Gallerande will then raise his siege to move to Lessines in order to join the army of the Marshal, who, to profit from the enemy's movement, should, I believe, move on Brussels."[523] As to Clermont-Gallerande, he explained to the Minister that, if it was necessary, he hoped on lifting the blockade of Ath and rallying on the camp at Alost.[524]

Fearing, no doubt, the consequences of d'Estrées' retreat and the probable results of the influence exercised by the Count and his friends, Maurice de Saxe had attempted, on 4 October, to reassure Clermont-Gallerande. He observed to him that the 15,000 men forming the maximum total of the Allied detachment and the forces that the Allies could draw from Mons, Namur, and Charleroi would not risk attacking the 22,000 under the command of Clermont-Gallerande, especially since d'Estrées had moved closer to him.[525] With the same

---

[522] La Tour to d'Argenson, camp before Ath, 4 October (Corresp. A.F.).

[523] Crancé to d'Argenson, Arbre, 5 October. (*Ibid.*).

[524] "A considerable enemy corps marched on Hal from where it detached a small corps to reconnoiter the camp of d'Estrées, which was at Enghien. A deserter, who had come to him from this detachment, informed d'Estrées that the Duke of Cumberland and von Waldeck were there with 10,000 men. D'Estrées then withdrew on Ghislenghien and the enemy detachment returned to Hal. In case that they marched in force on d'Estrées with a part of their army, d'Estrées after having received them shall retire on me. In case that they march on me, I shall retire on d'Estrées beyond the Dendre, and I shall consider returning my cannon to Tournai and marching diligently to rejoin the Marshal behind the Dender, which surely, at that time, will follow that they are united, shall be before Brussels and shall have cut all communication of the enemy with Antwerp. However, I do not believe that the enemy shall dare to risk this event, which is why after having established all my communications, after having taken all the necessary precautions in a parallel case, I shall vigorously continue my siege and not abandon it unless forced to do so or by orders from the Marshal." [Clermont-Gallerande to d'Argenson, camp before Ath, 4 October (Corresp. A.F.).

[525]Maurice de Saxe to Clermont-Gallerande, 4 October, Alost, 9 a.m., (Corresp. A.F.).

Maurice De Saxe's 1745 Campaign in Belgium

courier, Maurice addressed similar exhortations to Count d'Estrées.[526] These two letters had barely left when the Marshal received the letter in which Clermont-Gallerande republished the regrettable projects of unblocking Ath already expressed to d'Argenson. The lieutenant general added: "I hope that you will honor me with your orders in case you find something to change in the project that I have sent you." It became urgent to stop Clermont-Gallerande on the disastrous road that he appeared to be taking. Without delay, the Marshal sent him the formal order to continue the siege of Ath, which the Allies had no intention of disturbing. This injunction sufficed to cut short Clermont-Gallerande's projects. Divided, however, between the obligation to obey the Marshal and the pressing cries of d'Estrées showed him the certain danger of the coming Allied attack, Clermont-Gallerande began his preparation of the battlefield.

At 11:00 p.m., on the 4th, the battalions of the old English garrison of Ostend came out of Mons and camped at Gosselies.[527] From this moment, there remained nothing more for the Allies to do than to throw themselves into Mons under the protection of Howley and the four Austrian battalions destined to replace the English. As a result, resupplied on 6 October for 4 days, Howley prepared to fulfill this last part of his mission since d'Estrées, after having freed the road from Mons by Hal, Braine-le-Comte, and Soignes, was himself charged with achieving from the 1st.[528]

Howley moved on Soignes, on the 7th, completing in this manner half of the journey he was to cover.[529] This movement produced a general anxiety that reigned in the Ath camp, such that, on the 5th, Maurice once again attempted to reassure Clermont-Gallerande and d'Estrées.[530] It appears that these two had misinterpreted the arrival of the Austrians at Hal. Even so, the Allies did not send these troops until the same time when Howley found himself in a position to move on Mons. The French generals only saw in this movement of the Austrians, an effort to reinforce the detachment at Hal. By the additional work of French spies, who announced to Clermont-Gallerande, the departure of Howley, added that new forces would be replacing the

[526](*Ibid.*).
[527] Waldeck, 4-5 October
[528] *Ibid.,* 5-6-7 October.
[529] *Ibid.*
[530] Maurice de Saxe to d'Estrées and Clermont-Gallerande, Alost, 5 October, the morning (Corresp. A.F.).

troops that were about to leave. In the French camp, no one doubted that the entire Pragmatic Army would be marching and that the two detachments occupying Hal and Soignes only formed its advance guard.[531]

On the other hand, for several days emotions appeared to have carried away the Alost camp. Though he remained persuaded that the Allies were not seeking to break the Ath blockade, Maréchal de Saxe informed by a certain source of the arrival of the Austrian battalions in Hal had taken certain precautions. He had "opened a march in five columns from Ninove to Hal, to cause the enemy to fear that they were in danger of being cut off."[532] He also sent the Royal Household to Lessines as well as the Guards with 10 cannon to Grammont. This movement permitted Clermont-Gallerande to deploy 34 battalions and 79 squadrons.[533] "By means of these reinforcements," said Maurice, "Clermont-Gallerande shall be in a state to march against the enemy, if he comes too close, leaving d'Armentières, with two field battalions, 3 militia battalions, which de Brézé should bring up from Tournai, and the Royal Dragoons to continue the siege."[534] With an eye towards filling the holes produced by the departure of the Guard and the Royal Household, Maurice had called du Chayla to him, along with the troops that he held near Dendermonde.[535]

These measures did not suffice to calm concerns. The Marshal, however, continued to receive news that confirmed him in the idea that these Allied movements were solely to reassure the entrance of the garrison coming from Mons. "Despite the certainty of this act," he said, "I have told de Clermont to assemble and take up a position very close and in front of Ath and to send his baggage to Tournai. I have asked de Biron, de Montesson, and d'Estrées to do the same."[536]

This last disposition produced a result completely different from that anticipated by the Marshal. It did not reassure the spirits that had been agitated, since 2 October, by d'Estrées and his friends. They concluded that finally the Marshal had reported the imminence

---

[531] Maurice de Saxe to d'Argenson, Alost, 4,5,6, & 7 October. – "The enemies are marching on us in force; the rest of the their army has arrived at Soignes." [Crancé to d'Argenson, camp before Ath, 7 October, (*Ibid.*)].
[532] Maurice de Saxe to d'Argenson, Alost, 6 October (Corresp. A.F.).
[533] *Ibid.*
[534] *Ibid.* 7 October.
[535] *Ibid.* 8 October.
[536] Maurice de Saxe to d'Argenson, Alost, 8 October (Corresp. A.F.). – Biron and Montesson commanded the Guards and the Royal Household.

of a danger that so may clairvoyant eyes had seen for a long time. As a result, they feverishly hastened to face the peril.

"One worked softly," wrote the anonymous author, "on the preparation of the battlefield before Ath. On the 6th and 7th, the news had become more interesting by the march on Soignes.[537] One redoubled their work, having had news that they marched on our side, none of the generals still doubted that we would be attacked."[538]

At 5:00 a.m., on the 8th, 2,000 pioneers and as many workers completed the excavations that had been begun. One fortified a good position at Chièvres to form the right of an entrenchment formed in the area covered by the stream. Arbre was to flank the left of this sort of rampart. The dam at the Attre Bridge permitted the flooding of a mile of this terrain. The French then constructed some redoubts along the stream and a fortified Attre was linked to the Coucou Wood by an entrenchment running along the stream. The French raised a work destined to prevent the Allies from turning this wood. A lunette blocked the interval between the Coucou Wood and the Renard Wood. The northeast front of the latter was put in a state of defense. The defensive line extended as far as Melin-l'Évêque.[539]

At the same time, Crancé took up his dispositions in anticipation of the battle that was awaited.[540] "One repented, now, and quite

---

[537] R.A. – The news that the anonymous author references are those of the departure of the Ostend garrison and Howley's march on Soignes. The attention aroused that these movements could escape the vigilance of our agents. The spies indicated to the Marshal, on the 7th (Aulent to Maurice de Saxe, Philippeville, 7 October). – La Motte, commander at Philippeville, writes: "There have arrived, on the 5th, at Charleroi, around 4,000 men. They come from Mons from where they left at 11:00 p.m., on the 4th. They departed in the morning, at 6:00 a.m., from Charleroi, taking the road to Brussels. It is supposed that they are rejoining their army. I am assured that it is the old Ostend garrison." [La Motte to Maurice de Saxe, Philippeville, 6 October (Corresp. A.F.)].

[538] R.A.

[539] *Ibid.*

[540]"Since my letter of this morning, the Governor of Ath has raised the white flag at the time that we did not expect it; the King's Household and part of our troops had already marched to the Chièvres heights, where the battlefield had been studied.... In every case, I have undertaken, in all events, to act in concert with Foulon, Commissioner of War, for the necessary precautions that our hospitals are in a state to receive wounded and to have the quantity of wagons necessary to transport them. I have brought from Tournai the cooking pots and other tools that might be needed and I have asked of de Valicourt, who is employed here as a commissioner of war, to evacuate his hospitals to Lille in order to be in a state to receive all the wounded that we might send there." [Crancé to d'Argenson, camp before Ath, 8 October, noon (Corresp. A.F.)].

seriously, the wasted time and one endeavored to repair it," wrote the Count d'Estrées' friend. The French constructed bridges. Cannon were drawn from the trenches to fill the battle line.[541] The King's Household was put in the battle line so as to give the other troops, further away in the lines of circumvallation, the time necessary to occupy their posts, when Ath raised the white flag and immediately capitulated, which obliged or appeared to oblige the Allies to march in the Casteau Moors and from there to Mons."[542]

In effect, Howley left Soignes, on the 8th, "went," says von Waldeck simply, "to camp close to Mons and throw in this place the 4 Austrian battalions which had joined him at Hal." It was reserved to the particular ingeniousness of the Count d'Estrées and his friends to establish a relation between the capitulation of Ath and the movements of Howley.

The chronicler explains these events in the following terms: "Many people," he said, "did not doubt that the enemy was going to attack us. Seeing that he had not done so, they said, as it often usually happens, that they well knew that it was not the purpose of the Allies and that one had alarmed oneself too hastily. Maréchal de Saxe, who, as I presume, inwardly had trouble in forgiving himself his former inattention when he sent to Enghien a small corps that was unable to maintain itself instead of a large one that could have prevented any reason for concern, openly said that he had always reckoned that the enemy would not dare to attack us, and that they had not intended to do so. Apart from the fact that his letter to the Count absolutely proved the contrary; to wit, that one must not care, basing on such an unfounded opinion, to remain so quiet and to expose a besieging army to the uncomfortable obligation to interrupt such an important siege at a season that could make it more difficult every day , and to expose it to the uncertainty of a combat against a nearly equal corps, which can freely and safely withdraw, which is one of the main reasons for engaging in a combat in order to save a good fortress. De Clermont-Gallerande was also among those who claimed having seen through the intentions of the enemy, which was not, according to him, to raise the siege. The reason why he spoke thus was the inactivity in which he had remained, and for which he was reproached, which he felt and which he believed, might protect him from reproach. He

---

[541]One finds, in the *Correspondance*, A.F., the general order of battle of the army camped before Ath.
[542] R.A.

claimed with the greatest good faith that the enemy would never be so stupid as to divide themselves, lest the Marshal advance and attack [the forces] that had remained between Brussels and Antwerp. Hence, he concluded that he could only be attacked at Ghislenghien, because the Cambron Plain would remove the attacking corps too far from Brussels, and because he would not run the risk of moving so far away from that place lest he might not be able to join it. Consequently, he affirmed that he would not be attacked at all, because at Ghislenghien the enemy might well have no possibility to withdraw if the Marshal marched on Hal in order to cut him off and put him between crossfires. Reasoning like this, he took no appropriate measures so as not to be surprised. However, that utterly wrong reasoning could only be advanced to the level of Ghislenghien, which actually ran few risks. But nothing could be more wrong than the principle according to which it would be a mistake for the enemy to divide himself. They could have done it in the month of August, but in October, it was to be expected that the two separate corps were safe from insult in this season. If the Marshal had marched on Antwerp, he could not have taken any establishment there, the season being too advanced to execute a siege of this fortress, which was provided with a more than sufficient garrison, and the land was not tenable in the winter without being master of it. At the most they had to fear a sudden appearance [of the enemy] that would not last very long because of the feared rainfall which would be of too little consequence to keep us from undertaking it. One can say the same thing of the corps that marched on Ath, which could not fear moving on Mons to reinforce its garrison, already containing 10,000 men, and to be sure of retiring as it did and as it should do if it were defeated. One need not fear being besieged in this fortress, the season prohibited us from thinking of it. Besides what came to the Allies in these dispositions, even from a lost battle, they were not only assured of their retreat, but still to reunite by the highway from Mons to Brussels, which was practicable in all weather. Their march showed that they had reasoned in this manner. I have never, however, thought it certain that their project was to raise the siege of Ath. But it suffices that the thing was possible that one take precautions. They risked nothing in approaching, and it is certain that if we had remained in the position where we were, and of which they were not unaware, we would have played on them so well that it would have been necessary that they had lacked all firmness to undertake it. An army, without cannon, without communications, without an inundation to protect it, without lines, without a studied battlefield, without its right and left supported is not very formidable, above all having a siege to execute and an enemy before it, master of attacking where it will. Such was still our situation during the evening of the 7th; but such was not its situation on the 8th, because of the vigilance of the Count d'Estrées and the ardor that he inspired in the other generals to prepare to receive the enemy and profit from the advantages of the location that they had so well put to profit, that I doubt that the enemy after having studied them, would have considered attempting a battle on the 9th (because it required at least 48 hours to perfect the posts, which one regretted having lost when one learned that the enemy was no more than 3 miles from us, separated by the most beautiful plain in Flanders). I do not dare here to decide that the only goal should be, as one has said, to withdraw the Ostend garrison from Mons. The road that it took and could take even more surely before the siege of Ath, the day of our sojourn at Lessines, show that the Allies feared it."[543] However, it is much more to fear having marched such a large corps to cover it.

In sum, one sees that if the Marshal had conceived the plan exposed in totality to d'Argenson, on 23 September, d'Estrées, on his side, had established some other, built on projects given freely to the Allies. D'Estrées' mission especially consisted of assisting in all his activity the intentions of the Marshal, although they were different from his own. It does not seem that in this circumstance d'Estrées fulfilled the role that was reserved for him.

D'Estrées remained, therefore, responsible for the unjustified panic that reached its height, on 8 October, which would necessarily compromise the success of French operations. The emotion was born in Enghien, in d'Estrées' camp, where, barely arrived, this general officer saw himself "in the abyss between the Hal corps and the Mons garrison." Despite the exhortations of the Marshal, he prematurely retreated and upon his arrival at Ghieslenghien, he disrupted their morale by the unfortunate influence he exercised over Clermont-Gallerande.[544]

Since he had camped on the left bank of the Dender, the army lived on its accumulated provisions in the magazines at Alost, Ninove,

[543]Waldeck, 7,8, & 9 October.
[544]Cf. Waldeck, on the subject of the "promenade" of Clermont-Gallerande, p. 203. – One has seen in regard to the return of the Ostend garrison, the Allies, more interested in the affair than R.A., did not share the opinions of the chronicler.

and Ghent. These depots were initially only to contain the supplies necessary to feed the army until 25 September. Although they had been partially restocked, the extension of the campaign had nearly emptied them. In addition, the position of the camp did not favor the directions sent by Séchelles. With the difficulties that arose in the Hainaut, the army found itself far away in this region, which obliged the Marshal to make some examples to raise morale.[545] Finally, the region began to become exhausted and the bad weather was coming, such that Maurice made his dispositions to put the army into its winter quarters immediately after the fall of Ath. In addition, the Marshal had profited from the leisure given him by the siege to revise the preparations made for the first time in September, when he had projected leaving the army. He had ordered the modification of certain march orders and itineraries with an eye to unblocking Tournai and Lille. As the Allies occupied Mons, Tournai and Lille were, in effect, the necessary crossing points for troops destined to return to France by Flanders and the Hainaut.[546]

These last events had divided the French forces into two distinct groupings, of which Alost and Ath formed the respective centers. There were, as a result, two dislocations, the first for the troops of Clermont-Gallerande. On 9 October, the latter received orders from the Marshal.[547] In order to conform to them he immediately returned to Tournai the three militia battalions that de Brézé had sent up on 8 October. He quickly repaired the minor damage caused by the French cannon to the Ath fortification and destroyed the siege works. D'Armentières occupied Ath with three infantry battalions, a dragoon regiment, and the Grassins.[548] From the 10th to the 12th, the heavy siege

---

[545] The correspondence of Clermont-Gallerande with d'Argenson is completely impressed with the perplexity of this lieutenant general who was most hesitant between the advice of the Marshal and the exhortations of d'Estrées. It appears also that this latter did not act particularly cavalierly. "According to the conversations, d'Estrées brought everyone to his first opinion and the first observation that he threw over the terrain immediately decided him. The rest of the day and part of the following day were spent in discussions. De Clermont, who was not pressed, wished to see and fight these opinions on the location. He was there, on the 6th, in the morning and after having fought with all his forces, he was obliged to recognize that his juniors in service were not equal in the science of the trade and it was decided that one would not waste any more time in following d'Estrées' plan." (R.A.).

[546] Sergent, Commissioner of Wars, to d'Argenson, Alost, 15 September (Corresp. A.F.).

[547] Maurice de Saxe to Clermont-Gallerande, Alost, 9 October (*Ibid.*).

[548] Crémilles to d'Argenson, Alost, 9 October (Corresp. A.F.).

artillery left for Tournai, after which d'Armentières was provisioned with material and ammunition. Clermont-Gallerande retained only his field artillery.[549] At Ghent, Séchelles had been a bit surprised by the rapidity of the dislocation of the troops; but he had redoubled his activity so as to hinder nothing.[550]

On 13 October, the King's Household and the Guards left Ath for Tournai, and the Gendarmerie and the du Roi Infantry Regiment left for Condé.[551] The same day Rohan, Bettens, Orléans, and Andlau reached Tournai.[552] At the same time, Clermont, at the head of the remaining 22 battalions and 16 squadrons, accompanied by 20 Swedish guns, left the Lens plain for Beaumont. In this fortress, there was to be a definitive separation of the various elements of this column. Each of the elements was to take the road for its winter cantonments. Clermont-Gallerande then rejoined his headquarters at Audenarde, moving by Maubeuge, Valenciennes, and Tournai.

Obliged to pass in the proximity of Allied held fortresses to regain Beaumont, Clermont-Gallerande had to conserve sufficient forces to remain safe from attack until the moment when he had no reason to fear attack. He camped, on the 13th, at Escaussines.[553] He reached Erquelines on the 14th.[554] He remained there on the 15th and 16th, sending his artillery to Maubeuge under a strong escort.[555] On the 17th finally, he reached Beaumont. His mission was completed. "It was still de Clermont who led these troops as he already led the others. The march which it made, on the 13th, below Mons, passed by the Saint-Denis Abbey. The crews and the infantry arrived during

---

[549] Crémilles to d'Argenson, camp before Ath, 9,10, 11, & 12 October. (*Ibid.*).

[550] Séchelles to d'Argenson, Ghent, 9 October (*Ibid.*). On the 10th, the Marshal reported to the Minister that he had given his orders for the dislocation without much care for the presence of Howley's 8,000 men still camped at Saint-Symphérien, below Mons. This latter waited there, no doubt, for the movement of the French on Mons, dreaded by the Allies.

[551] The du Roi Regiment reached Valenciennes and the Gendarmerie the Champagne. The guards left for Paris on the 16th.

[552] Bettin was to remain and on the 14th, the rest went into their quarters.

[553] The troops formed into two columns. That of the right contained, in the following order, 4 companies of grenadiers, camping detachments, the Picardie Regiment, 3 field guns, the cavalry (4 regiments), 3 guns, and the Auvergne Regiment. That of the left was for Seedorf, the rest of the artillery, the baggage, and the old guard. The vicinity of Mons had suggested this formation.

[554]The march was done in the same formation as the day before.

[555] This escort was a battalion of the Hainaut Regiment destined for Landrecies and the Anjou, Brancas, and Bourbon Regiments, containing 12 squadrons. Maubeuge found itself on the road of quarters reserved for this cavalry.

the night and continued arriving until 8:00 a.m., the following day."[556]
This is based on the impression that the friend of the Count d'Estrées
left de Clermont-Gallerande.

During this time Maréchal de Saxe had completed the preparations for the dislocation of the army which he had decided not to
march out until after Clermont-Gallerande had moved away. Following the definitive project established, all the elements of the army were
to remain united until after the crossing of the Escaut over the bridges
at Audenarde, Ghent, and Gavre. In Gavre, where there was no fixed
bridge, the French had constructed two with the army's bridging train.
Arriving on the left bank of the Escaut, the different corps prepared,
finally, to separate.

On 9 October, the 9 battalions and 8 squadrons were brought
below Alost under the same conditions that du Chayla had marched
to Dendermonde. The French transported everything that remained in
the magazines at Alost and Ninove to Ghent. On the 12th, the French
constructed the two bridges at Gavre and finally, on the 13th, the army's heavy baggage preceded the troops, leaving for the future camp
at Baerleghem.

On the 14th, the army marched out in its turn. Formed in four
columns, it began its movement at 7:00 a.m. It marched in an inverse
order, "the right of the camp forming the left of the march." In all
the columns, the cavalry took the head, followed by the infantry. The
two center columns head brought five Swedish guns with them and
the two wings each had 10 guns. This artillery marched with the two
last brigades of infantry forming the tail of each column. The right
wing contained more cavalry than that of the left because Allied detachments could come out of Antwerp and trouble the march. For the
same reason, the Eu-Crillon Brigade, camped below Dendermonde
departed at 6:00 a.m., marched directly on Ghent as a flank guard.

---

[556] "As the weather is bad and that most of the soldiers had no tents, de Clermont had
judged it proper to have them canton until the day when they could depart successively to reach their garrisons. I have come here before the camping detachments to reg-
ulate said cantonments with de Sainte-Marie, Commissioner of War, to whom I have
given a rendezvous. I have sent de Clermont, before the army, to give each troop the
state of its cantonments where they had all arrived before 2:00 p.m. I have regulated,
with de Sainte-Marie, the states for the provision of forage, wood, and straw. I have
sent the army's butchers who will furnish the troops until the day of their departure
and they will draw bread from Maubeuge." [Crancé to d'Argenson, Beaumont, 16
October (Corresp. A.F.)].

Henry Pichat

Each column received a specific rearguard, which did not prevent the formation of two strong general rearguards, each provided with 10 field guns.

Because of a misunderstanding of the usual rules, the troops were to camp at Baerleghem without observing the order of battle. They installed themselves without regard to seniority. One observed only that each of them was within range of its debouches to march the following day. In order to avoid disorder, the different detachments of artillery had parked, during the evening of the 13th, at the head of the infantry brigades with which they were to march the following day. The baggage wagons that marched out at 1:00 a.m., on the 14th, had swept the terrain. One used the road network as they normally did. However, the French avoided using the highway to Ghent which was not considered very secure.[557]

The 14th, at midnight, a brigade of artillery, with 10 light guns, left the Baerleghem camp with a good escort. It crossed the Escaut over the bridge farthest upstream from Grave. This artillery took a position so as to cover the left of the march that the army was going to make on the 15th. It was also to cover the lifting of the pontoons from the Grave which was done immediately after the passage of the columns.

At daybreak on the 15th, the French troops left their camp at Baerleghem formed in four "divisions" the first three of which formed one column and the fourth forming two. The first division crossed the Escaut at Audenarde and camped at Waemaerde. The second and third divisions, respectively crossed the two bridges at Grave and occupied the villages in the immediate vicinity of Audenarde. The rearguard of the third division lifted the pontoons which a company of artisans then took to Ghent. Besides, after completing this operation, the brigade of 10 light guns also went to the same city, where they were to be reassembled with all the French field artillery.

The fourth division crossed the Escaut at Ghent. Its right column was formed of cavalry, followed by infantry with 10 Swedish

---

[557] The movement order for 14 October (Corresp. A.F.). – The general rearguards placed, respectively, under the command of a lieutenant general or maréchal de camp, that left that day were important. That which covered the two right hand columns consisted of the old guards and the posts of the left camp, 6 piquets, 10 companies of grenadiers, 10 cannon, the Carabiniers and the Saxe Volontaires (in total 14 squadrons). The other was formed of the old guards, and right hand posts, 6 piquets, 6 grenadier companies, 10 guns, 100 cavalry carabiniers, 100 dragoons and the Beausobre Hussars.

cannon and marched in the same conditions as the day before. The other column consisted of the Carabiniers, the baggage wagons of the corps of this division, and those of the headquarters. The two columns joined at Ghent. The rest of the artillery not employed in this column, but which traveled with it, waited at the Baerleghem camp, which the baggage wagons, of which we will speak, had to defile. The artillery was to follow the baggage and park with it on the glacis of the Saint-Lieuvin Gate. This convoy went into city behind the right column, formed with the only troops designated to remain in Ghent. The troops of the left column having orders to canton in the villages in the neighboring region passed through the city first.

Each division received a rearguard and the heavy baggage of each left at 3:00 a.m., before the troops. The assistant baggage masters led the army from Baerleghem to Audenarde, to Ghent and to the two bridges at Gavre. At these various crossing points over the Escaut, the columns found local guides requisitioned to lead them to their respective camps and cantonments.[558] These two marches of the 14th and 15th were done in perfect order and most rapidly.[559]

Finally, on the 16th, all the elements of the army set out to move to their winter quarters. Before the departure, the command assured itself that the baggage as well as the escorts and the rearguards knew exactly the location of their destination. The relocation was completed in complete tranquility. From the beginning of this movement the Allies did nothing.

According to the state of the quarters established by the Court, the troops were divided into 11 commands linked to the headquarters at Ghent where the Maréchal de Saxe was to remain. These subdivisions echeloned as follows: 2 on the Dender (Ath and Dendermonde),

---

[558] All the details of these last marches are summarized in the movement order for the 15th, transcribed by Brézé.

[559] Crémilles to d'Argenson, Ghent, 19 October (Corresp. A.F.). – "The troops entered Ghent and the arrangement was made with much facility because of the precautions that had been taken in advance. The cavalry, which is in the quarters, will be there some days for the necessary dispositions to be made, but everything shall be done. There is nothing left behind and all the oats and rye that we had in Alost and Ninove are now in our magazines in Ghent. I have already returned the mobile hospital to Lille and I have disbanded all the unnecessary employees and surgeons. I have done the same for the mail, the constabulary, and the mounted constabulary. I shall have the honor to send to you a detailed report on these issues. I have been, in addition, pressed to reduce expenditures. It was not an easy task, but a prompt debacle, above all that the weather is terrible, constantly raining for the last three days. [Séchelles to d'Argenson, Ghent, 15 October (Corresp. A.F.)].

4 on the Escaut (Ghent, from Ghent to Tournai, Tournai, la Scarpe, and the upper Escaut), 1 between Sambre and the Meuse, 2 on the Lys (Courtrai, la Deule), and 2 in Maritime Flanders (Ostend, Dunkirk). This establishment realized the complete achievement of the program laid out by the Maréchal de Saxe in the *Mémoire* of December.[560]

The separation of the Pragmatic army followed shortly. On 9 October, the Allies received word of the French preparations and their reconnaissances confirmed its execution. On 19 October, the Allied troops went to their respective quarters, covered by the Senne to the West and to the south, by the Haine, the Sambre, and the Meuse. Sixty-nine battalions, 98 squadrons, and 12 free companies, formed the total of the Allied forces wintering before the French. Among the Allied forces there were only 18 English squadrons of the former 29, as much cavalry as dragoons, in the Pragmatic Army on 29 June. As for the British infantry, it had completely vanished from the Lowlands.

The commentary which finishes the writings of the anonymous chronicler and zealot of d'Estrées naturally found their place here. "It is thus," concludes the writer, "that the campaign finished that one can call the most beautiful campaign that was ever fought in Flanders. One sees by these memoirs that it was an enterprise without a plan, led without prudence, successful by accident, to which war does not always respond happily. The principal event and that which to which the campaign owed its success was due to the skill of d'Estrées, the valor of the troops, and to the lack of wisdom by the enemies. That which followed, with the exception of the surprise of Ghent, did not appear sufficiently organized or conforming with the true interests of France to do as much honor to the general who had appeared, during all their course, who lead from day to day without organized plans, such as one had in the best campaigns of Louis XIV led by de Turenne and de Luxembourg. The resolution taken from the month of July did

---

[560]De Vault and Brézé. – The latter furnished, in truth, a little different distribution. It mentions, in effect, a 12th command formed with a certain number of troops united at Dunkirk and Boulogne, under the eventual command of the Duke de Richelieu, with an eye towards a landing in England which, one knows, never occurred. It was to consist of 13 battalions and 9 squadrons. In addition, Brézé placed de Vaux, commander at Dendermonde, under the authority of Du Chayla, commander at Ghent. As for the total effectives, Brézé had 162 squadrons and 117 battalions, which came to 78 line and 39 militia battalions.

It is necessary to consider that the Army of Flanders was sent to canton in the Evêchés, the Thiérache, and the Champagne with Clermont-Gallerande's troopsl that is to say 22 battalions and 50 squadrons. These troops do not figure in Brézé's total.

not appear to have been imitated in the future. The people who did not like long campaigns nor to campaign late thought that de Belle-Isle, to whom they attributed of having caused the Court to execute it, acted only to thwart the Maréchal de Saxe and said openly that he had risked the glory of the campaign for a pointless siege. These people knew neither the land nor the necessity of occupying the important post of Ath in the time when one could, with the certainty of succeeding, and they saw only the blame for the risk that one ran would fall on the generals who executed it and not on those who had advised them. Ordering the siege of Ath, was to order its capture with security and precaution, and it was necessary only to send 20,000 men to Enghien, who would easily subsist there and now draw on the magazines on the Dender, which were beginning to empty out. In spite of that this last event did not contradict the others, and one can admire this campaign it was as as admirable as it was inimitable."

This conclusion contains a singular mixture of different feelings. It seems however to establish that the anonymous writer and his friends do not decide to give their whole approval to Maréchal de Saxe. In their eyes Maurice was, above all, a lucky general. However, he was no longer the intruder formerly so badly greeted by the general officers of the French army. At the end of 1744 one reproached him for having left a French province "to be eaten."[561] At this hour still without recognizing his talents, one compares him to Luxembourg, and even Turenne. Royal favor and popular acclaim have consecrated the reputation of the Marshal; but his own collaborators did not wish to consider the victory of Fontenoy and the captures of Tournai, Audenarde, Ghent, Dendermonde, Bruges, Ostend, Nieuport, and Ath, in six months of campaign, as the result of fortunate events. Everyone, however, is not of this opinion. De Brézé comes to another conclusion which he shares with us: "One has rarely seen" he said "so I believe, the end of a campaign from which an army separated in such a good state is ours after a battle and seven sieges. I will agree that the majority of these sieges were not considerable. But uncertain if the attacked places would not defend themselves as far as they could, it was necessary to make all the necessary preparations, so that once attacking them, one would be sure of the success of the enterprise. This would not be without great fatigues for the infantry. In spite of that, however, it left this campaign in a better state than it had after less arduous campaigns." Coming events of considerable importance were soon going to emphasize all the force of this last consideration.

---

[561]Colin, *Les Campagnes du mar*échal *de Saxe* (loc. cit.), Vol. II, p. 595.

CHAPTER V
The Preface to the 1746 Campaign in Flanders.

LA TOUR D'AUVERGNE. — GATINAIS. — BLÉSOIS.
En 1746.

On 12 October 1745 there appeared a "Regulation" for the lodging of French troops in Flanders in their winter quarters. It contained the following dispositions:

1. He who commands the King's Army shall be lodged where he judges appropriate and he shall be furnished for his house furniture and utensils, fire and candles, as he orders.

2. A lieutenant general shall be lodged and furnished the furniture and utensils or 4,000 florins at his choice for winter quarters (a florin shall be evaluated at 36 sols, 8 deniers, of French currency). A maréchal de camp shall have 2,000 florins, a brigadier shall have 1,000 florins. The Maréchal de logis de l'Armée [Chief-of-Staff of the Army], the major general of the infantry, the *maréchal de logis de la cavalerie* [chief-of-staff of the cavalry] shall be treated according to their grades as maréchaux de camp or brigadiers, when they do not hold the rank of brigadier. The *aides-maréchaux généraux des logis* [assistant chiefs-of-staff] or *aides-majors généraux* [adjutant generals] shall equally be treated as brigadiers. The Chief Engineer shall re-

ceive the treatment of a lieutenant colonel of infantry, at least, if there is not a superior rank and thee other engineers as captains of infantry, at least if there is not a superior rank. The artillery officers shall be treated according to their ranks. The staff officers of the fortress shall also be treated according to their ranks. The captain of guides who shall be employed at Ghent by the maréchal [de Saxe] as well as the Baggage Master General shall be treated as infantry captains and the captain of guides shall be given a place to lodge his guides. The quartermaster shall be treated as an infantry lieutenant. The maréchal's aides-de-camps, if he judges it proper to keep them near his person, shall have for lodging and utensils attributed to their rank. The *commissaire ordonnateurs* shall receive 800 florins for lodging and tools and the *commissaries ordinaries* shall each receive 500 florins.

The lodging of cavalry, infantry, and dragoon officers shall be marked in the same quarter where their regiments are established. If they find the means to have it, by mutual agreement, the city shall pay accordingly: to a colonel of infantry, cavalry, or dragoons, by month, 40 florins; to a lieutenant colonel of cavalry – 30 florins; to a lieutenant colonel of infantry – 25 florins; to an infantry or cavalry major – 25 florins; captain or adjudant of cavalry or dragoons – 20; an infantry lieutenant – 9; a cornette – 10 florins; to a sous- lieutenant of infantry or Enseigne, priest, or surgeon major – 6 florins.

3. If the lodging is given in kind with furniture and tools, nothing shall be paid and the lodging shall be as set forth in the *Regulation of 2 April 1711*.

4. If one gives, as lodging, an empty house, without furniture and utensils, the general officers and others shall receive half of that which is indicated in the foregoing articles and, in case the officer, does not find lodging by mutual agreement, the magistrates will give him an appropriate lodging in the quarter with his troops, according to his grade, and at a price below the sum fixed for his housing.

5. There shall be furnished, by the land, the quantity of wood and lighting listed in the following schedule to each of the general officers and others: to a lieutenant general, per day, 100 logs, 1 sack of charcoal, and 5 pounds of candles; to a maréchal de camp - 60 logs and 3 pounds of candles per day, and 8 sacks of charcoal per month; for a brigadier – 50 logs and 2 pounds of candles per day, and 15 sacks of charcoal per month; to a cavalry, infantry, or dragoon colonel – 12 logs per day, to a lieutenant colonel – 10; to a major – 8; to a captain

Henry Pichat

or aide-major – 6 logs; each lieutenant – 4; each sous-lieutenant,, Enseigne, cornette, priest, or surgeon major – 3 logs; to a *commissaire ordonnateurs* 60 logs and 3 pounds of candles per day, and 15 sacks of charcoal per month and to the other commissaries 50 logs and 2 pounds of candles per day, and 8 sacks of charcoal per month. This allocation shall only be issued to actual and present officers.

6. In the case where the general officers prefer to furnish their own wood and lighting, it shall be bought back at the rate of 12 French livres per 100 logs, 5 livres for the candles. The magistrates shall be free to buy back the wood that may be furnished to the officers at the following rate: each colonel 24 florins; lieutenant colonel – 20; major – 16; captains or adjutants – 12; lieutenants – 8; sous-lieutenants, cornettes, enseignes, priests, and surgeon-majors – 6. In this case then there shall be established a magazine from which an officer may purchase wood at a price which shall be regulated and not that of the repurchase.

7. No general officer or other officer shall take a lodging that is of a quality appropriate for a higher-ranking officer.

8. No general officer or other officer shall profit from winter quarters except in proportion to the time that he remains at his post, so long as he is not employed elsewhere by order.

9. The winter quarters lodging shall not begin until the day that the troops enter and shall continue only until they depart it, and it shall only be furnished for those men actually present.

10. The magistrates shall pay great attention that the barracks or other lodgings destined for the lodging of soldiers, cavaliers, or dragoons, are in good condition and the materials are provided according to regulation, that is: one ration for 3 soldiers and one for 2 cavaliers or dragoons, if the beds are of a size conforming to the King's ordinance, if not they shall be supplied.

11. The coal shall be given to the soldiers during the winter at the rate of 1½ livre per man per day, and one shall furnish the firewood necessary for lighting. There shall be furnished to each guard corps, which exceeds 30 men, 150 livres of coal per 24 hours and 4 candles (of the rate of 11 per livre). One shall provide the guard officer's room 70 pounds of coal per day, or charcoal in proportion, if there is no chimney, and 4 candles per day. When the guard corps does not have 30 men or is of an inferior number, they shall be provided with only 70 livres of coal; but one will always give them 4 candles. One shall give

4 candles (of the rate of 11 per livre) for every groom of 50 horses or fewer, and 8 candles for grooms with more than 50 horses.

12. Nothing shall be demanded at the gates of the city by those who enter or who leave, be it in money or kind, such as wood, faggots, grain, or other foodstuffs.

13. All officers and soldiers shall be obliged to pay the taxes, excise taxes, or other charges of the provinces and cities without being able to claim any exemptions by means of the canteens which will be established in all the cities.

14. There shall be given to the magistrates a state of the employees handling food, forage, and hospital for the service of the army to which the lodging must be furnished according to their functions and who, in no case, shall receive no buy-back, nor provision of wood or lighting.

One sees that this regulation appears to create a particularly favorable regime which the troops would enjoy at the end of October. The Intendant Séchelles presented, however, some doubts in this regard.

On 1 November, the Minister fixed the composition of the staffs and designated the general officers employed in the winter quarters, 9 lieutenant generals and 13 maréchaux de camp received letters of service which gave them the right to the salary corresponding to their rank and employment. [562] On the other hand, no brigadiers were employed. Those who, by their personal choice, remained with the troops, received the right to a forage allocation. The minster promised that he would request to the King that these latter be accorded, after the winter, a gratification proportional to their services.[563] No staff was officially constituted. The Marshal had requested to conserve with him, at Ghent, 12 officers, among whom were his ordinary collaborators and favorites.[564] The Minister authorized them, but none of these officers received service letters. They were promised the same gratifications as the brigadiers. As for the staffs in the conquered fortresses, their situation remained definitively regulated as the Marshal had provisionally established it upon the moment of the occupation

---

[562]State of general officers that the Maréchal de Saxe proposed to employ during the winter on the frontier of the Lowlands (Corresp. A.F.).

[563] D'Argenson to Maurice de Saxe, Fontainebleau, 20 October (Corresp. A.F.).

[564] State of the staff officers proposed to the Count d'Argenson to remain with me during the winter [Maurice de Saxe to d'Argenson, Ghent, 26 October (Corresp. A.F.)].

of Ath.[565] In sum by the different measures one avoided, above all, creating privileges in favor of the general officers and the staffs of the troops wintering over in Flanders to not be obliged to join the Armies of Germany or Italy and, thus, multiply expenses.[566]

To lodge the soldiers, the French first used the barracks in the conquered fortresses. These buildings had not suffered. It was necessary to distribute all the officers and part of the troops with the inhabitants and indemnities of their lodging constituted a heavy charge on the region. The citizens of Ghent, provided with an important garrison, constituted a special fund in light of dealing with these dispenses. It was necessary to provide these funds by means of a voluntary contribution raised by subscriptions to avoid the cost of lodging which would be entirely provided by municipal revenues. Audenarde, Nieuport, Ostend, and Dendermonde, with lower populations and less rich than Ghent, could not imitate Ghent in this. Their magistrates presented loud complaints to Séchelles. It was unjust, they claimed, that they were no longer allowed their ancient privileges to be exempt from the lodging of soldiers since His Most Christian Majesty had, upon taking possession of their cities, confirmed and maintained all their rights.[567] The intendant transmitted these complaints to d'Argenson, adding that he propose to authorize the municipalities to withdraw the funds necessary for the utensils of the French troops "at the expense of newly conquered Flanders, when it shall be established."[568] D'Argenson responded with a firm "no." "The King did not want that expense to be admitted to compensate for any grants, nor did he want it to reduce its yield."[569]

Other protests were raised. It is known that from the beginning of the month of August, the Brussels States had authorized the populations of the Brabant to execute Séchelles' orders. The magis-

---

565   "With regard to the staff in the conquered fortress, I form them with the officers of the garrison that find themselves in each of these cities, having it directed by the lieutenant-colonel or the most senior functions of the King's lieutenant, the functions of a major by those of the garrison reputed to be the most intelligent and so on. There shall be formed a mass of money, which will be enjoyed by those who are employed by the Austrian Government. The individual who commands the greater part and the other shall be divided on a pro rata basis to those employed as a King's lieutenant, as a major and as adjutants." [Maurice de Saxe to d'Argenson, Alost, 13 October (*Ibid.*)].

566 D'Argenson to Maurice de Saxe, Fontainebleau, 20 October (Corresp.A.F.).

567 Séchelles to d'Argenson, Lille, 18 November (*Ibid.*).

568 Séchelles to d'Argenson, Lille, 18 November (*Ibid.*).

569 D'Argenson to Séchelles, Versailles, 1 December (*Ibid.*).

trates of the region requested, now, that one deduct from the sum of their impositions the foodstuff's that they had provided to date.[570] The intendant did not accept this request. He remarked that these orders had preserved the region from a requisition, pure and simple, and as a result, it was impossible, by submitting to the request of the magistrates, to pay again for these supplies by an unspecified writing-off of the impositions.

It is appropriate to consider, and it is there, no doubt, that one finds the successive causes of the amount due, which still remained due to the French Treasury, at this moment, of nearly 3,000,000 livres, the sum of the impositions made on Flanders for the operations of 1743, 1744, and 1745.[571] A single exception was made: Marchault, Intendant of the French Hainault, received orders to not request of Austrian Hainaut and of the County of Namur anything more than was strictly necessary, "in such a manner as to diminish as much as possible the funds contributed by this region."[572]

It soon appeared that Séchelles had exaggerated nothing in presenting some reserves on the favorable results that one generally expected from the regime accorded to the French troops.[573] Because of the importance of the garrisons of these cities, the growth in the consumption had become more rapid and noticeable during the quartering than it had been during the campaign. The rising of the prices of foodstuffs had immediately responded to this augmentation. During the course of his inspection tour of these cantonments at the end of October, Séchelles received the complaints of the troops.[574] Everywhere the soldiers found life too expensive. For the numerous French cavalry, in particular, the situation became very difficult. Except on the campaign they were obliged to supply themselves from the magazines of the French intendancy, under more onerous conditions than if, wintering in France, they had drawn forage from the depots of their garrisons. Moreover, losses caused by an extremely disadvantageous exchange were still added to the cost of food. In these conditions, the army would quickly find itself in a situation as bad as Belle-Isle in Bohemia during the winter of 1742-1743. The allocations accord-

[570]D'Argenson to Séchelles, Versailles, 5 December (*Ibid.*).

571   The precise sum was 2,884,481 livres, 4 sols, 7 deniers [Receipt made on the arrears of contributions, Séchelles to d'Argenson, Lille, 17 December (Corresp.A.F.)].

[572] D'Argenson to Séchelles, Versailles, 5 December (*Ibid.*).

[573] Séchelles to d'Argenson, Ghent, 9 November (*Ibid.*).

[574]Séchelles to d'Argenson, Bruges, 3 November (*Ibid.*).

ed to the troops were barely sufficient to live on. There were insufficient funds to put the regiments in condition before the opening of the next campaign. In view of fixing this unfortunate state of things, the King granted the Army of Flanders an extraordinary gratification of "good food" of 2,000,000 livres. Maréchal de Saxe received orders to distribute this sum in concert with Séchelles, the inspectors-generals of the infantry and of the cavalry, as well as "such other persons as they judged useful to consult." The Minister ordered "to distribute the gratification by including there, following the express intention of His Majesty, that only the officers who took part in the provision of utensils, and particularly those charged with the repairs of the troops, and to also establish what could be given of this sum to the soldiers, cavaliers, or dragoons." No other officer of the troops or of the staff could receive the least part of this gratification. The division ordered to the headquarters of Ghent was followed almost immediately with the distribution at the beginning of December.

Upon their arrival in the conquered fortresses, the French troops completed the repair of the defenses. The work has been begun almost the same day as the capitulations, but no part of them was completed. The barracks and houses that the French cannon were damaged were repaired. In Ath and Ostend, in particular, nearly 800 houses were raised.[575]

Everywhere the troops were regularly exercised in conformance with the Marshal's orders, who directed also that the strictest discipline be observed. Otherwise, all the infantry corps commanders received formal and precise instructions from the Minister of War.[576]

---

[575] "When I entered Ostend, there were 837 houses destroyed or very damaged where the repairs required at least 531,733 florins, according to the estimates of the workers and I did not wish to undertake them at that price, which came to, in French currency, 974,885 livres, 16 sols, 8 deniers. When I left (4 November) they had already been repaired at an expense of 350,509 florins, 10 sols, which were, in French currency, 642,601 lives, 15 sols, more than 700 toises of pavement at 17 sols, 6 deniers, two barracks structures at the expense of the city which cost, in French currency, 14,500 livres, two ovens constructed anew for the commissary officers, at a cost of 2,200 livres; which came to a total of 665,425 livres, 15 sols." [State furnished by Guignard *Commissaire ordinaire des guerres* employed at Ostend, 15 November (Corresp. A.F.)]. – To this report were attached the nominal states of the property owners whose houses were rebuilt in Ath and Ostend.

[576] This circular dated in November is held in the correspondence of A.F. (Supplementary Carton 1745). This dead letter does not remain; because in the many reports addressed in November and December by the governors of the fortresses to d'Argenson and the Marshal have frequently found a report that "the troops have exercised."

"The King wishes that his troops be exercised during this winter," he said, "with more care and regularity than in the previous year. His Majesty has ordered me to order you that his intention is that as soon as the regiment that you command shall be established in the garrison or the quarter designated for it, you will profit from every favorable day to assemble it with the flags and have it execute the manual of arms, marches, evolutions, and movements in use, putting, as much as possible, no gaps in these exercises; above all for the recruited soldiers that it is necessary to have maneuver under cover when the weather does not permit the assembly of the regiment. His Majesty has given his orders that there be delivered to his troops sufficient powder that they may frequently fire [their weapons] and that it be augmented to the measure that the corps shall be reinforced by recruits. Finally, he wishes that there be nothing spared that the soldiers be exercised as it is appropriate for the advantage of his service. It is necessary, principally, to teach him to manage his fire, that he be accustomed to fire appropriately, and that he commit himself to redressing the fault, into which one frequently falls, of firing simultaneously the fire of all the ranks. His Majesty wishes absolutely to be informed each week of the progress of these exercises and of the quantity of times that they have been exercised during the week, with the reasons for which they were missed. You can be assured that I shall have great care to report to him the exactitude that you bring to this effort."

"On our new frontier, there was nothing to fear except that the training of the troops should not slow. The necessity of watching over the preparation of the various contributions and also that of preserving the populations from reprisals or exactions by the enemy shall oblige the commanders of the fortresses to exercise an active surveillance. The enemy shows himself initially very enterprising, above all in the parts of Austrian Hainaut and the County of Namur occupied by French troops. His emissaries spread false edicts by Louis XV that are very vexing to provoke the emigration of the peasants. De Machault has imprisoned the scandalmongers who spread these ordnances."[577]

The Allies pillaged French convoys. They also terrorized the mayors of the region, forbidding them from obeying the orders of the intendant, and they raised contributions from the French side of the border. This manner of acting was absolutely the activity of the 12 free companies that the Allies had distributed in Nivelles, Vilvorde, and

---

[577] Correspondence of Machault and d'Argenson in October and November (Corresp. A.F.).

along the road running along the Brussels canal, since the dispersion of their army. Waldeck had attached to them 1,000 Bavarian hussars cantoned in Brussels, Jemappes, and Nivelles, the horsemen in the Mons garrison, as well as the Ferret and Béthune partisans gathered in one place. All these troops formed mobile detachments ready to pillage.

The vigilance of d'Armentières, Le Danois, and Phélippes put an end to the audacity of this horde. They vigorously chased the Allied detachments that ventured south into French territory. They sent emissaries to requisition the mayors who complained anew to the intendant. Marchault ordered the imprisoning of these imprudent messengers, and they did not return. However, the Allies still used the mail to send demands to the municipalities. It became urgent to cut short these maneuvers to reassure the populations. "People of the countryside are impressed by this," sent Marchault to minister; "I have written a circular letter to the directors of the offices of my department to stop these letters and send them to me." D'Argenson approved, adding that he agreed "to hold these orders secret as this precaution not do any tort [wrong, injury] to the postal system by destroying the confidence of the public."[578] Besides, one took serious precautions. D'Armentières, commander of Ath, deployed the foot and mounted Grassins. He installed a post of these troops at Leuze and another at Frêne-sur-Ronne to assure communications with Tournai. Beaumont was put in a state of defense.[579] One established a garrison there consisting of a battalion, five companies of militia, and a squadron.

If the communications between Philippeville and Meubeuge were henceforth assured, they still remained precarious between that city and Valenciennes. By the means of the ancient Roman road between Mons and Cateau and favored by the covered ground, the Allied detachments coming from Mons reached Bavay. This latter intersection of numerous roads led them there where they decided to pillage the town. The blow struck, the Allied pillagers returned to Mons rapidly and easily. The only thing that remained for them was to deter-

---

[578]Correspondence of Machault and d'Argenson in October and November (Corresp. A.F.).

[579]"To erect parapets, the old walls were used, which were lowered in some places and raised in others, in order to put the soldiers under cover and in a state to fire while standing and along the towers attached to the walls. The intervals between them were pallisaded in order that the Allies could not quickly reach the foot of the walls. Three quarters of the wall were sharply sloped such that one could regard this post safe from attack and in a state to oblige an enemy to attack with cannon.

mine, based on information from their spies, an opportunity of another such exploit. In the hope of catching these pillagers, Le Danois and Phélippes had organized daily a reconnaissance that left Valenciennes, Maubeuge, Avesnes, and le Quesnoy.[580] Believing these measures insufficient, Phélippes proposed to the Marshal that they occupy Bavay. As this locality was in the region commanded by Le Danois, he was given this task. It was executed on 20 November. The Maubeuge Director of Fortifications immediately went to Bavay, putting it into a state of defense.[581]

The works were not completed by the time the Allies attempted to attack this new post. On 24 November, at 10:00 a.m., the partisan Béthune left Mons with 300 hussars and dragoons, supported by 200 infantry. He put the infantry in ambush in Trainières, which he hoped to burn. Le Serre, commander of this tiny fort, came out with his troops at the first news of Béthune's approach. For two hours a violent battle occurred, after which Béthune abandoned the fight and left many dead on the battlefield. The Allied hussars then pillaged along the road to Valenciennes and so that they would not be pointlessly disturbed, they devastated several French hamlets. The mail coach from Maubeuge arrived in the middle of this brawl. Béthune's horsemen pillaged it, caring little that it traveled with a passport signed by the Queen of Hungary. The Allies then retired on Saint-Ghislaine and Mons with booty estimated at more than 6,000 livres, "all of which belonged to the poor people."[582] These details prove that it had become indispensable to put a brake on the aggressiveness of the pillagers. Hitherto the French governors had only sent out patrols that had no effect. They sent after each exaction protestations to de Nava, the Austrian commander at Mons. These remained without effect. The complaints also remained purely platonic because de Nava greeted them with the most perfect courtesy and perfect indifference. From

---

[580]Correspondence of Phélippes and Le Danois with d'Argenson during the months of October and November (Corresp. A.F.).

[581] The three gates to the city were closed and pallisaded redoubts were constructed before them that were capable of housing 50 men. In addition, in the interior, a redoubt was constructed where one could retire in case of need and as much as possible a parapet was constructed." (Vault, p. 115, note 1). – These summary works sufficed to put Bavay in a state of defense from a coup de main. The Mons garrison, with its 11 battalions and 300 horse, could not attempt an expedition against Bavay without provoking a response from the troops in Valenciennes, Maubeuge, le Quesnoy, and Avesnes upon the first alert.

[582] De la Serre to d'Argenson, Bavay, 25 November, at 10:00 p.m. (Corresp. A.F.).

the occupation of Bavay, a nearly complete security reigned in that region, however.

The Allied incursions, however, between Alost and Dendermonde, as well as along the Escaut, were much less disturbing. Von Waldeck's Bavarian hussars could barely move into the region where the de Beausobre light horsemen and the Saxe Volunteer Uhlans patrolled in force and frequently. Besides, the Marshal had recommended to avoid giving a too vigorous pursuit of the Allied incursions. He wished to give them confidence "so as to give them a good washing [shellacking] at the first occasion."[583] In sum, the Dender, which covered the front of the French winter quarters, proved a barrier that was not easy to cross.

To the north, finally, Löwendahl, commander at Ostend, proposed to Maréchal de Saxe to occupy Damme, whose fort should constitute an advance post on the Dutch maritime fortresses along the Ecluse and the Sas-de-Gand. This position could no longer, with utility, cover the city of Bruges, which was covered only by a wall. The Marshal approved this project which, nonetheless, remained unexecuted because it was recognized that it was impossible to house 60 men in Fort de Damme.

Since the dispersion of the Pragmatic Army, the spies and governors of the fortresses did not gather enough information to completely develop an exact situation of the Allies' quarters. The English and Hanoverian troops moved without cease. An embarkation of a large part of the British infantry had preceded the dispersion of the Allies. It was soon proven that there was no longer a single English soldier in Flanders and that the London Court sought to recall a part of its mounted troops. They could, however, only embark with their arms and equipment, their horses being left behind at Antwerp.[584] The departure of the infantry for England and the movement of the British cavalry towards Antwerp provoked continual movements of the troops which made it difficult to establish an exact situation of the Allies' quarters.

The Austrians, on their side, reinforced Luxembourg with troops drawn from the garrisons of Mons, Namur, and Charleroi. Kau-

[583] For reasons that will become apparent later, the Marshal also desired that the populations of these regions become accustomed to seeing French light cavalry. The presence of von Waldeck's hussars deprived Maurice the opportunity to too vigorously pursue the Allies.

[584] Reports and letters from Brussels, 6-22 November. – de Saxe to d'Argenson, Ghent, 4 December (Corresp. A.F.).

nitz, in effect, was excited to see arrive in Lorraine the 22 battalions that the Marshal had sent there without prejudice to that which Conti caused to move to this frontier after having separated from the Army of Germany on 2 December.[585]

Thus, since the end of the campaign, the Pragmatic Army was no longer definitively installed in its winter quarters. One could not see at which time the Allied troops would cease to shuttle between the various places. The Duke of Cumberland had left for England where he would take command of the troops operating against the Pretender. The old Marshal von Koenigsegg, sick and fatigued, had returned to Vienna.[586] Only Waldeck assured the command of the Lowlands.

Since the Allies had established themselves on the right bank of the Senne, the Prince had perfected the defenses of Brussels as well

---

[585] "France has moved not only from Flanders but also several other places many troops into Lorraine and on the frontiers of that region. It has made their dispositions and preparations from which one can truly infer a siege of Luxembourg is its objective. [Van der Duyn (commander of the Dutch troops in Brussels) to Kaunitz, Brussels, 20 December (Corresp. G.H.). – "Not to await that an extremely active enemy take from the Allies the means of holding Luxembourg." Kaunitz gave the order to reinforce this fortress [Report addressed, on 22 December to Lieutenant General van der Duyn (Corresp. G.H.)]. – For a long time the Minister had already received word that the Allies were actively working at Luxembourg. "There have been for 3 years in Luxembourg 400 artisans who work continually on cutting the rocks that serve as the foundation for several fortifications and works of this fortress. There is at the foot of the bridge that leads to Mansfeld Castle an underground passage where there are three vaults that are bomb proof. They can contain 4,000 men. The magazines are filled and there are provisions of every type in great number. The officers and the bourgeois expect that they will be bombarded in the winter and besieged in the springtime. Von Neuperg had declared to the bourgeois that in case of a siege the city would be reduced to ashes before he would surrender it. Desertion [among the garrison] was considerable, many soldiers breaking their arms and legs by jumping into the ditches. It contained, on 1 November, 8,000 men. They expected 7,000 more consisting of cavalry, infantry and pandours." [Joyeuse de Grandpré, Mézières, 3 November (Corresp. A.F.)].

[586] "The fatigues of these first marches of this campaign have already shown me how greatly reduced my forces are and I was no longer in a state to remain on horse after the battle of Fontenoy. The gout which struck me a few days later has exhausted me and for more than for months, I have found myself immobile, unable to walk without the support of an arm, nor to remain on a horse without being greatly pained in the knees and kidneys. It is a sad task to make war in this state when one can barely hope that a man of my age, 73 years, and a body exhausted by 52 years of unrelenting service. An army is very bad in the hands of a man who can no longer act by himself and who is obliged to trust reports that are frequently late, and equivocating, and finally to run the risks of making blunders and irreparable errors above all on the day of a battle." [Koenigsegg to Their High Powers, Beaulieu, 22 September (Corresp. G.H.)].

Henry Pichat

as those of Antwerp and fortified the works constructed on the Vil-vorde Canal. However, there was a rumor that Waldeck was about to receive a reinforcement of 10,000 men from Germany and that one prepared to lodge them in the religious houses and communities of Brussels.

This rumor received no credence. One did not see, for the moment, at least, from where these troops could come. The Repub-lic had sent 6,000 men to England, without prejudicing the effectives that they held in Flanders and Germany. Could the Dutch troops in Germany come back to the Lowlands, when since 9 November, Conti was always in observation on the left of the Rhine, still occupying the lines of the Queich, to disperse its army [into winter quarters] only on 2 December? On the other hand, one did not believe that the Austri-ans would withdraw, at this time, a single grenadier from the army of Prince Charles of Lorraine. Frederick II [of Prussia], recently victori-ous at Soor, was knocking at the gates of Lusace, threatening Dresden and Bohemia. The Hessians and Hanoverians, maintained in the Prag-matic Army, respectively, 6 battalions, 9 squadrons and 6 battalions, 16 squadrons. Did they find themselves in a condition to provide new contingents? The troops that they could raise would do nothing more than replace the British battalions that had moved to England. In sum, some circles or certain Electors of the Empire, seemed, alone, to be able to provide these 10,000 men. No word from the French ministers announced that these German Princes appeared disposed to espouse the quarrels of Maria Theresa, even though they had made her hus-band Emperor. It was more probable that if the embarkations for the troops for England continued, the Dutch would soon be the only forc-es protecting their territory.

From the beginning of November, the Court resolved to profit from this situation to exercise the reprisals that it had considered from some time regarding the Republic. The dispersal of the French army, after the capture of Ath, had only delayed the hostile projects that it had formed. By virtue of the indulgence of Clermont-Gallerande, the Court had still not found the occasion to execute them. It now pre-pared to resume them.

D'Argenson brought the Marshal up to date with the Roy-al Instructions. His Majesty, wrote the Minister, had resolved to not declare war on the Republic, at least for the moment. He proposed, however, to unexpectedly open hostilities against Holland by some important operation prepared with care, considering "that it should be pointless to strike before one was in a state to strike."[587] However, the Court did not wish to prematurely fatigue the troops before their return on campaign. It counted; finally, on leaving them time to rebuild themselves. Did these various resolutions permit it to consider major enterprises? The season of the heavy rains which fell at this time was not propitious. Communications were now only possible by the main paved roads. It was thought it could undertake later, in the heart of winter and over the heavy ice, a sudden attack against l'Écluse, Sas-de-Gand, or some of the other neighboring Dutch fortresses at the head of the French quarters. This project appeared reasonable. One hoped, at Versailles, "that the States General, always thrifty, had been capable of neglecting to add to the natural defenses that they had so well fortified."[588]

The Marshal requested some delay to respond to these overtures. He had still not been able to procure the plans for these fortresses. Nonetheless, information taken, he had to abandon the illusions that he had forged. No doubt, the Dutch held themselves on their guard. The uproar caused in August by the approach of French had not yet calmed. Everywhere the Dutch had broken the dams and spread the flooding. If, by virtue of a frost, the French could approach the above mentioned cities, they would find themselves facing solid works and batteries that had been recently established. The Marshal estimated that one could not consider doing more than insult these fortresses. Would such a little result be worth deranging the troops?[589] One sought something else. In awaiting, the Court seized with eagerness the opportunity to satisfy a desire expressed by the Marshal. Maurice had complained several times already that conquered Flanders remained open to the many Dutch officers or soldiers who came to travel there or pass their semester leaves there. This tolerance facilitated espionage and the soliciting of deserters. The Minister authorized the Marshal to close the land, but he added a sharp restriction to the powers given to the Maréchal de Saxe for the circumstances.[590]

---

[587] D'Argenson to Maurice de Saxe, Fontainebleau, 10 November (Coresp. A.F.).

[588] *Ibid.*

[589] Maurice de Saxe to d'Argenson, Ghent, 18 November (Coresp. A.F.).

[590] "His Majesty desires that you write a letter to this commander of Dutch Flanders by which you indicate to him that although Holland is not in a declared war with France, it appears to you that it was your prudence to not suffer on the King's lands any soldier serving the States General, who actually served in the troops of his enemies; that it was a purely a precaution of war and that should not influence the commerce that is

It seemed that Court did not manage to find the occasions which it sought. It then considered the project of occupying the Island of Cadsant. D'Argenson consulted Folard on this subject. [591] The old chevalier had, in effect, penetrated onto the island during the month of May 1708, by passing between Écluse and Sas-de-Gand. He affirmed that it could be easily maintained if the Duke de Vendôme had shown some vague desire to support him. As in the winter one could not think of slipping between the two cited fortresses, Folard remembered the use of special boats, called "prames" to invade the Island of Cadsant.[592] D'Argenson greeted, without enthusiasm, Folard's very detailed propositions, estimating that in sum this latter had only furnished him "with vague and general observations, on which one could not form a fixed and solid plan of operations."[593] Besides, this project did not interest the Marshal any more. Maurice thought that towards the north the French could not obtain results sufficiently important to compensate for the cost of the expedition. Why then, however, did one construct these useless boats? "There are good things to draw from Chevalier Folard," said Maurice; "but frequently his imagination stimulates him and carries him too far."[594]

Since he was in possession of desirable information on all the Allied fortresses in the Lowlands and the establishment of the Allied army, Maréchal de Saxe shared with the Minister a project that he had conceived. He proposed the capture of Saint-Ghislain, adding that the possession of this fortress favored an attack on Mons while the frost was hard. Maurice affirmed that the preparations for this expedition were nearly already complete. The capture of Saint-Ghislain however

---

done freely between the two states and that your plan is not to interrupt nor trouble; that, finally, he must inform the troops that are in the service of the Republic that this sort of permission will no longer be requested and that you will grant no more of them. You will easily conceive, by that which I shall have the honor of sharing with you, that the King does not judge that it is appropriate that the orders that you will give on this subject appear to have emanated from His Majesty; but that it is only because of the reasons of war which one makes you take the step to prohibit all these comings and goings as being able to allow espionage and reconnaissances that you wish to prevent. As for the rest, His Majesty exhorts to you to make carry out this order with the least possible commotion, not that He is concerned with sparing the Republic at any point; but because when he wants to indicate his dissatisfaction with it, it should not be by such weak means."

[591] D'Argenson to Folard, Fontainebleau, 15 November (Corresp. A.F.).

[592] Folard to d'Argenson, Paris, 2 December (*Ibid.*).

[593] D'Argenson to Maurice de Saxe, Versailles, 27 November (*Ibid.*).

[594] Maurice de Saxe to d'Argenson, Ghent, 5 December (*Ibid.*).

constituted only the least great part of that overall plan of which the Marshal traced out the broad outlines. "There ran through Brussels the rumor that the Hessians and Hanoverians must pass to England. If they arrive, I shall be of the opinion that we should leave our quarters to take Brussels and Antwerp. I believe that the Dutch will evacuate the rest of the Lowlands without being told twice.[595]

The Hessians and Hanoverians had, in effect, left Louvain, Malines, Diest, Aerschot, and Herenthals where they had established themselves upon the dispersion of the Pragmatic Army. As for the English cavalry, they had moved closer to Antwerp.[596] As a result the garrison of Brussels remained isolated and the fortifications of this place were not sufficiently powerful to long sustain an energetic siege. Finally, if the French attacked Brussels suddenly, this city would not receive any assistance from Mons, threatened itself, by the capture of Saint-Ghislain.

The Marshal's propositions left, on 5 December, for Versailles and it appeared that they would be warmly greeted. They received, however, an unexpected response. They arrived a bit late at the Court and, above all, at the time when other patriots occupied their minds. On 13 December d'Argenson sent to the general headquarters at Ghent the state of the troops that the King had resolved to send to England. Nineteen battalions and 9 squadrons formed part of the forces thus designated and two other regiments of cavalry were to follow the first movement. [597] Thus, the French assembled at Calais the provisions of every type necessary for the expedition. The Duke de Richelieu appeared to have decided to press the embarkation of troops. In awaiting to go shortly to Ghent in order to coordinate with Maurice, the Duke had d'Argenson to ask him to make some demonstrations that would favor the enterprise against England.[598] As these preparations immobilized part of the French troops in Flanders, the Minister thought that simple demonstrations alone remained possible in that region. He

---

[595] (*Ibid.*).
[596] Maurice de Saxe to d'Argenson, Ghent, 4 December (*Ibid.*).
[597] D'Argenson to Maurice de Saxe, Versailles, 10,11,12, and 13 December (Corresp. A.F.). – State of the troops that should embark, Versailles, 11 December (*Ibid.*).
[598] The Marshal received the invitation under the form of a report attached to the previously cited letter of 13 December. It appeared, no doubt, to the Minister that this missive would not be sufficient to decide Maurice de Saxe to give all his assistance to Richelieu and to his enterprise. One reads, in effect, in this report, that the Marshal "is too good a citizen to not lend himself to it and too great a general to indicate to him only what had to be done."

proposed to Maurice that he obliged the Allies to leave in the Low-lands the Hessians and Hanoverians by means of some simple and skillful operation. Otherwise, sincere or not, d'Argenson affected now to only vaguely believe in the coming embarkation of the Hessians and Hanoverians.[599]

According to the comments of this same minister, on the sub-ject of Dutch Flanders "things had gone well" since the beginning of November "which would seem to form new obstacles with the project that His Majesty wished executed."[600] For reasons that do not enter into the framework of this study, thus do not justify being examined, the king "had taken to heart" the cause of Charles Stuart.[601] The Court anticipated the embarkation of Richelieu for 20 or 25 December at the latest. It proposed to the Marshal supporting the enterprise with a small diversion; but "it shall only be done in the case where he judges his movement would not damage the rebuilding of the troops.[602]

No matter what it was and in spite of these appearances, Maurice was not to be long in realizing that the last proposals sent to the Minister had been actually taken into serious consideration. Fur-thermore, the Marshal did not give up any of his projects and while waiting for Richelieu's arrival at Ghent he did not cease to show an absolute lack of confidence in the success of the preparations being pursued with so much ardor in Calais. The meeting that he had with Richelieu soon confirmed him in this assumption.[603] Maurice came to the point and removed some illusions from Richelieu who did not show, it seems, a great knowledge of the details of the [military] trade. "He thought," reported the Marshal, "among other details, that the cavalry had been loaded like sheep.[604]

Irrespective of the fate reserved for the expedition to England, Maurice thought that if the French troops in Flanders were to brave the rigors of the weather to undertake a demonstration, it should be at least logical to draw a practical result from the material losses that such an operation would necessitate. He decided, as a result, that he would change nothing of the plan he presented, on 5 December, to d'Argenson. He resolved, only, to perfect the preparations of his en-

---

[599]Correspondence de d'Argenson with Maurice de Saxe in December.

[600] D'Argenson to Maurice de Saxe, Versailles, 13 December (Corresp. A.F.)

[601] *Ibid.*

[602] *Ibid.*

[603] Richelieu arrived at Ghent during the evening of 27 December.

[604] Maurice de Saxe to d'Argenson, Ghent, 28 December (Corresp. A.F.).

terprise in taking into account the new difficulties that were imposed on him by Richelieu's expedition. The Ghent headquarters therefore took the cleverest steps to remedy the weakness of the means of action imposed on them. It became, initially, indispensable to calm the Allies' concerns raised by the continuing presence of Maurice in the middle of the winter and in contrast to the normal practices of the time. In this regard, historians have given us a luxury of details, so we will not discuss them here.

The necessity to protect their preparations from the indiscretions of "the court chatterboxes" was real. It appeared, however, that it was pushed a bit far because Versailles was unaware of anything. This was the cause of the first cruel disappointment for the Marshal.

At the end of November and to reassure von Waldeck of his intentions, Maurice had sent a great number of his colonels on leave. Without preparatory warning, on 17 December, d'Argenson informed the Marshal that the King had accepted the representations of the Duke de Richelieu on the necessity of making a demonstration in Flanders. As a result, the Minister officially returned to the army all the infantry, cavalry, and dragoon colonels Maurice had sent on leave. This order, however, did not require any colonels other than those in Paris to return. "The intention of His Majesty was not that any disruption be produced in the progress of recruiting or the rebuilding of the units."[605] It was, in effect, a true demonstration that the Minister was to execute. One with such "brilliance" in the month of December would indicate nothing less than a complete resumption of hostilities. Unfortunately, in the actual state of affairs, no measure was more likely to render useless all the precautions with which Maurice had surrounded his preparations as the moment of execution approached. The Marshal complained bitterly: "You have asked me to execute demonstrations," he wrote d'Argenson, "in preference to a real operation, however you have acted in such a way as to remove any chance of success in its execution. All my arrangements were made with de Brézé and other general officers for the beginning of operations on the 28th, without them knowing anything more than when they should arrive, not having given any of them their objective. Finally, I had hoped, and not without some basis, on giving the King a present and as proof of my re-establishment, the capture of Brussels with 10-12 Dutch battalions, that of Saint-Ghislain and several quarters of the enemy captured, without having risked anything; because they have no general, no assembly point, no dis-

[605] D'Argenson to Maurice de Saxe, Versailles, 17 December (Corresp. A.F.).

positions, nor preparations to make any, and once I have moved into the middle of their quarters I shall have nearly a month before me to maneuver and profit from the opportunities that war and fortune give me. I have not dared to announce to you such great things because it is better to do than to say. I had counted, however, on sending you my plan around the 24th to receive your approval. There would have been, then, time to dispatch the colonels, if you had judged it proper. Pardon me, sir, the chagrin that I feel for this setback.[606] Thus the complete development of the plan summarily exposed by the Marshall appeared on 5 December. Von Waldeck had returned to La Haye. The assembly of the Hessians and the Hanoverians, in the region of Antwerp, was completed. Certain Austrian troops from Mons, Charleroi, and Namur had moved to Luxembourg. The Allies had not received any reinforcement from Germany. Under these conditions, the Marshal could count on surprising the Allies in an absolute disarray in presenting himself suddenly before Brussels. To be certain, the city would be carried and nothing would then oppose the penetration of the French into the heart of the Allies' winter quarters.

It appears that this letter re-awoke the interest of the Court in favor of the Army of Flanders. On 25 December, d'Argenson responded that the King had approved everything and thought "that the movement for its execution was very close."[607] Success was not doubted. It was only regretted that Maurice considered the future occupation of Brussels as useless, even dangerous. D'Argenson recommended that he dismantle its defenses so that the city would remain completely open and useless to the Allies. He thought, finally, that the disruption that would be made was not irreparable, because immediately after Richelieu's embarkation, the Minister proposed to return the colonels to their leave. "One pushed back the enemy into greater safety, on side of Lowlands, where they were not before, so England was re-awakening."[608] For a little while and thanks to an inversion as unexpected as singular, Richelieu was now going to play the role which 15 days before had been reserved for the Marshal.

Thinking "that it was not by the posting of the enterprises that surprises become easy" Maurice had suspended his movement.

[606] Maurice de Saxe to d'Argenson, Ghent, 20 December (Corresp. A.F.). – This letter produced a response from d'Argenson, written on the 23rd, in which the Minister excused himself for upsetting Maurice's projects, blaming the ignorance in which Maurice had left him.

[607] D'Argenson to Maurice de Saxe, Versailles, 25 December (*Ibid.*)

[608] D'Argenson to Maurice de Saxe, Versailles, 25 December (Corresp. A.F.).

[609] "Your courier came at the right moment; he informed d'Argenson; the troops were to receive their different orders tomorrow, and they are already in the hands of the general officers, for which I am very annoyed." [610] At Douai one stopped the dispatch of the prepared artillery material, which had already begun. The requisitioned horses and wagons were going to be returned to their owners and those of the artillery reserve returned to their stables and sheds. Everything was to be put back in place. One continued to remain in complete silence. One only sent the ladders and beams requested for the expedition against Saint-Ghislain. [611] Otherwise, the thaw that had begun on 24 December and it was feared that the water would turn to rain. This was the capital reason for delaying the march on Brussels for some more time, because one could not count on hard frosts before the end of the period of the rains, which, according to the peasants, did not still appear about to end. The Marshal decided that only the expedition against Saint-Ghislain should occur. [612] The preparations were completed and the first project had been addressed to the Court. In its great lines, this plan anticipated the use of four cannon "which one would dispatch two days earlier on the road from Douai to Beaumont, to have then, by a counter order, at the point named at Valenciennes." [613] The French formed two attack columns. The first moved along the right bank of the Haine and contained 300 grenadiers and 400 fusiliers "for the surprise" supported by 300 horse and 2,000 infantry. The second, formed like the preceding, operated on the right bank of the river. The forts of Ath, Valenciennes, Condé, and Maubeuge were to furnish the necessary troops. The garrison of Saint-Ghislain contained no more than 700 men, such that the French forces employed sufficed to contain the defenders of Mons as well. [614]

On 24 December, in the course of an interview with Maurice, de Ramsault, engineer to Condé, spoke out against the project addressed to the Court, affirming that he considered the project as impractical, even during the frost. However, knowing perfectly these locations, Ramsault led the French troops to Saint-Ghislain with the assistance of his son, by moving down the Haine in a number of boats

---

[609] D'Argenson to Maurice de Saxe, Versailles, 20 December (*Ibid.*).

[610] D'Argenson to Maurice de Saxe, Versailles, 25 December (*Ibid.*).

[611] Saint-Périer to Maurice de Saxe, Douai, 26 December (*Ibid.*).

[612] Maurice de Saxe to d'Argenson, Ghent, 25 December (*Ibid.*).

[613] Project for the surprise of Saint-Ghislain (*Ibid.*).

[614] *Ibid.*

Henry Pichat

previously assembled at Jemappes or at Quaregnon. He proposed to launch an attack, at the same time, on the exterior of the ramparts of Saint-Ghislain, counting that supported by this surprise, the garrison would have neither the time nor the means to resist.[615] It was important only that the Governor of Mons have no suspicion, because the preparations proceeded in the greatest secrecy and the attack would be made before daybreak. The idea appealed to the Marshal as it provided him the means of creating a diversion favorable to Richelieu's expedition at the time as the Court had fixed. Ramsault was finally filled with ardor. "He told me that he responded positively on the success of the project," wrote Maurice to d'Argenson and better informed by his experience, he added "these sorts of things are so subject to chance that I do not dare to answer you about it."[616]

The project of the engineer having been adopted, in principle, the Marshal sent it back to Le Danois and Phélippes, leaving them to develop it with Ramsault. "They are closer than I to receive the information; they are in a better state to judge it," thought Maurice.

Charged with the organization of the operation, on 25 December Le Danois and Phélippes abruptly decided, on the 29th, to inform Maurice de Saxe that the expedition against Saint-Ghislain should occur at night.[617] That same day, at daybreak, Le Danois dispatched from Valenciennes 550 men for Condé. These soldiers crossed the Haine there. Then moving up the river by Ernisart and Pomerouel, they reached Baudour and Tertre at nightfall. They moved into ambush in the woods near the small villages in such a manner as to mask Mons. This troop was not to come out until they heard the sound of the attack, moving against the Saint-Ghislain Gate, which was the closest, and to make a diversion there. Le Danois left Valenciennes with the main body of his column, towards 4:00 p.m., while Phélippes, obliged to move down a longer road, left Maubeuge at 3:00 p.m. The junction of these two lieutenant generals was to occur at 1:00 a.m., in Quaregnon.[618] They strongly occupied Jemappes and embarked about 1,000 soldiers with Ramsault and his sons. At the price of some efforts arranged in advance, the engineer would penetrate into Saint-Ghislain. He would "have the French march sounded" when he judged it appropriate. At this signal, the 6,000 men under Phélippes

---

[615] Project for the surprise of Saint-Ghislain in Hainaut (*Ibid.*).
[616] Maurice de Saxe to d'Argenson, Ghent, 28 December (Corresp. A.F.).
[617] *Ibid.*
[618] Le Danois to Maurice de Saxe, Valenciennes, 34 & 30 December (*Ibid.*).

and Le Danois would attack the fortress's ramparts.[619] Moreover, the success of the enterprise especially depended on Ramsault's success, because Phélippes and Le Danois could not hope to force the guarded redoubts, the covered way and palisades as well as the walls flanked with crenelated towers surrounding Saint-Ghislain.

The garrisons of Valenciennes, Maubeuge, Bavay, and le Cateau had furnished the necessary troops. In truth, Le Danois could have deferred the expedition a few more days; the preparation for the surprise would no doubt have gained had he done so. However, as the Governor of Mons had loaded the available boats at Quaregnon and Jemappes with charcoal, it was feared that boats would be lacking, and it was resolved not to wait until they were all employed. In addition, the rumor that one was about to attempt a *coup de main* on Saint-Ghislain had spread throughout Paris. "I have received a letter by which one informed me of this," said Le Danois. This indiscretion decided him to hasten.

Everything failed miserably. Delayed by the difficulty of the roads, Le Danois did not reach Quaregnon until 5:00 a.m. Phélippes was not at the rendezvous. He was awaited vainly until 8:00 a.m. Not thinking himself sufficiently strong to undertake the enterprise in full daylight, Le Danois retired on Quiévrain. He had barely entered this region, towards 9:00 a.m., when Phélippes arrived in his turn. Leaving Meubeuge at the agreed-upon hour, this general officer continued his march through the darkness of the night and was led astray by misinformed or ill-intended guides.

These guides had led him into the middle of some swamps from which his troops could only escape after having climbed through ditches, thrown bridges and constructed paths of fascines. The expedition had failed.[620] As for the engineer, he wrote to the Marshal, not without some irony: "As the operation, in what concerned me, was to start only in Quaregnon and on its shore, which is still a half-mile away, these gentlemen will give you a report of what they have done and ordered."[621]

This little check did not discourage him. "This failed blow," he added, "does not prevent me from flattering myself that I can still

---

[619] Project for the surprise of Saint-Ghilsain en Hainaut, *loc. cit.*

[620] Phélippes to Maurice de Saxe, Maubeuge, 31 December; Le Danois to Maurice de Saxe, Valenciennes, 31 December; Maurice de Saxe to d'Argenson, Ghent, evening of 31 December, (Corresp. A.F.).

[621] De Ramsault to Maurice de Saxe, Condé, 31 December (*Ibid.*)

find the means to surprise Saint-Ghislain, if you or the Court desire it. In this case I shall execute another project where I will detail what I believe necessary to observe, as much from the departure from our fortresses as on the march. These details will not wound anybody because nothing will be issued but your orders without me saying a word."[622]

These lines denounced with wonder the causes of this failure. They established also the responsibilities for each participant. Otherwise, the Marshal estimated that Le Danois had made a first fault "by pushing [forward] a detachment of 500 men in broad daylight and passing close to Saint-Ghislain."[623] Le Danois made a second error in omitting to reconnoiter the roads that he had to cover during the night and during a period of full thaw. For the same reasons Phélippes had obtained an even worse result. Whatever it may be, the attack on Saint-Ghislain failed. The complete result of that which the Marshal prepared to direct in person on Brussels became ample compensation for this failure.

On 28 December, the thaw became complete seeming to have delayed for a long time the march on Brussels. "I am very delighted to have suspended our operation," said Maurice, "because of the Senne and the Vilvorde Canal, which became null when they froze, removing all difficulties."[624]

Towards 15 January 1746 the freeze resumed, but the Marshal waited until the ice was sufficiently solid to cross. On the 18th, a second thaw ruined these growing hopes. "If it continues," said Maurice to the Minister, "I shall return all the colonels.[625] When I have the honor to see you, I will speak to you about the means of arriving at our goal, but one needs a full field operation for that."[626]

Otherwise, there circulated news that was not very favorable. It was said that the Hessians and Hanoverians would receive orders to remain in the Lowlands. One also spoke of the pending arrival of Allied reinforcements from Germany in the Brabant.[627] If in the month of November 1745, this last rumor was unlikely, one could now give it some credence. The Prince de Conti had, in effect, dispersed his army on 2 December 1745. In addition, Frederick II had completely aban-

---

[622] *Ibid.*

[623] Maurice de Saxe to d'Argenson, 31 December (Corresp. A.F.).

[624] Maurice de Saxe to d'Argenson, Ghent, 28 December (*Ibid.*)

[625] D'Argenson had given, on 1 January 1746, complete latitude to the Marshal in this regard.

[626] Maurice de Saxe to d'Argenson, Ghent, 28 December (Corresp. A.F.).

[627] Maurice de Saxe to d'Argenson, Ghent, 20 January (Corresp. A.F.)

Maurice De Saxe's 1745 Campaign in Belgium
doned his alliance with France, having signed the treaty of Dresden, on 26 December 1745, with all France's enemies. "There is a chance of frost, especially from now to the 15th of the coming month," said the Marshal who abandoned none of his projects, "I will profit from this if I can find a moment, because it is only a question of crossing [the river], as the means of return are not difficult."[628] In Versailles, they were anxious, even impatient. The situation of affairs in Europe remained sufficiently distinct that one thought of preparing a campaign plan for 1746 if one wished to open hostilities early. The King awaited Maréchal de Saxe, at Versailles, for the definitive elaboration of a program of operations.

On 22 January, the freeze resumed, provoked by a very violent north wind. So great was his impatience, the Marshal did not change his habitual circumstances. His preparations continued as the people of the country gave him certain reassuring indications.[629] In reality, Maurice had taken a definitive decision and for the little that d'Argenson knew of his character and habits, he could not doubt, after receiving this last letter, that an expedition was about to occur.

At first the forecasts of a heavy freeze given by the peasants appeared about to be realized. On the 24th, the boat bearing the daily courier from Ghent to Bruges had to stop half way because of the ice.[630] However, two days later a thick fog appeared. On the 27th, the sky was cloudy, it rained, and the temperature rose.[631] This proved the complete and definitive ruin of these last hopes. It was also for the Marshal to prove that he had attempted the adventure. Upon the announcement of the thaw, he sent the last orders to begin the march. Had he not written a dozen years before in his *Rêveries*: "In war it is frequently necessary to act on inspiration. Circumstances feel better when they are not explained and if war holds inspiration, one should not disturb the soothsayer." Maurice could barely hope to find a better opportunity to join an example to the precept. Neither waiting, nor procrastinations were to be found in the tenacity of the Marshal. It appeared that his desire to act was still more firm by the numerous proofs than the circumstances imposed on him since 5 December. The capture of Brussels would serve as a preface for the 1745 campaign; at Versailles, one awaited the arrival of the Marshal with less impatience.

---

[628] Maurice de Saxe to d'Argenson, Ghent, 20 January (Corresp. A.F.)

[629] Maurice de Saxe to d'Argenson, Ghent, 23 January (*Ibid.*).

[630] Maurice de Saxe to d'Argenson, Ghent, 25 January (*Ibid.*).

[631] Maurice de Saxe to d'Argenson, Ghent, 27 January (*Ibid.*).

During this time, the preparations for the expedition were pursued in the greatest secrecy. The Allies' suspicions had subsided at the end of November. The unfortunate expedition against Saint-Ghislain had reawakened them.[632] Their concerns were calmed again when the Allies lightly reinforced Mons, Namur, and Charleroi where important provisions were accumulated. According to the information provided by de Courbuisson, Commander of Bruges, Kaunitz, whose alarm had been very high at the end of December, now appeared to believe that the French would make no further movements during the rest of the winter.[633] Otherwise it was generally thought in Flanders that the French desired peace above all since the King of Prussia had abandoned his alliance with France.

However, von Waldeck remained most distrustful. His horsemen frequently pushed reconnaissances towards the Dender. Almost every day they "squabbled" with the Marshal's uhlans and hussars who did not hesitate to engage them, in order that the Allies become accustomed to see them in the region of Brussels. On 16 January 1746, returned from La Haye to Antwerp because there was a rumor that the French were about to make a movement.[634] He gathered all the generals present with the Allied army. It was decided that on the first alert those of the troops that still occupied Antwerp, Malines, Louvain, and the vicinity would immediately come out to take up a position on the right of the canal running from Brussels to Vilvorde from Antwerp to Brussels. A part of the garrison of this latter city would join these forces so as to reform the line established earlier by the Allies in the

---

[632]On 29 November, in a conference held by Kaunitz, the Allies had taken some precautions to cover their winter quarters. They were thought sufficient after Kaunitz had stated that "though it is to be hoped that one would not be disturbed before the beginning of the next campaign, the possible enterprises of the enemy during the winter had appeared to him to require the generals to gather and deliberate together on the precautions that it was prudent to take to be prepared for any event as well as on the manner of carrying out them out in case of need." [Waldeck to Their High Powers, Brussels, 2 December (Corresp. G.H.)]. In reality the progressive departure of the Hessians and Hanoverians upset these projects, as it also rendered useless those that von Waldeck would form a bit later. It is necessary to say that the Allies did have sufficient time to suspect the Marshal's plans.

[633] Courbuisson to d'Argenson, Bruges, 30 January (Corresp. A.F.). The Prince of Hesse had indicated, on 26 December, that there was some agitation in Ath. At this time, von Kaunitz showed himself particularly disconcerted. [Van der Duyu, Commander of Dutch Troops in Holland, to Prince von Waldeck, Brussels, 28 December (Corresp. G.H.)].

[634] Waldeck, 16 January 16 January 1746.

month of August 1745. Their plans established, von Waldeck returned to La Haye to report to the States General.

Towards 26 January 1746 the Allies were considering embarking new troops for England. Four Hessian battalions left Antwerp for Wilhelmstadt. Three Hanoverian regiments, with their artillery, were to fill the gaps thus produced in the Antwerp garrison.[635] This movement produced the complete evacuation of Malines and Louvain. Under these conditions the precautions taken by Waldeck became illusionary because the arrival in Flanders of Imperial reinforcements, as certain as it was, was not, however, occurring soon. The clairvoyance of Maréchal de Saxe had not been wrong. The moment to act that he had known to await had arrived.

On 27 January, the interested general officers received the firm order to set out "on the dates fixed for all the points from where they were to leave."[636] The different preliminary concentrations were completed on the 27th. The movements that they had necessitated were not long. The troops chosen to form the six divisions, which were to march on Brussels formed the major part of the garrisons of Maubeuge, Ath, Tournai, Audenarde, Ghent, and Dendermonde or cantoned in the vicinity of these fortresses from which the six divisions were to depart.[637]

Two divisions, commanded respectively by d'Armentières and Clermont-Gallerande, were to invest Brussels from the Upper Senne. The third, under the orders of de Vaux, prepared the passage over the Vilvorde canal and the Lower Senne for the fourth division brought from Ghent, by the Marshal himself, to isolate Brussels by the north. De Brézé brought, from Tournai, the siege artillery escorted by some troops forming the fifth division. As for the sixth division, placed under the command of de Phélippes, it was to take up a position at Binch to observe Mons and Namur.

The march and the composition of each division was set as follows: d'Armentières left Ath during the night of 27/28 January with 4 battalions and 7 squadrons. He marched on Nivelles and, on the 29th, moved between Brussels and Louvain, leaving the Soignes Forest to

---

[635] La Graulet to d'Argenson, Ghent, 27 January (Corresp. A.F.).

[636] *Ibid.*

[637] It is necessary to add to the concentrated movements of the troops forming the six divisions those that were necessary to replace all the forces that left these garrisons. One did not wish to leave them stripped of troops. The gaps were filled, by preference, with militia battalions.

its left. Clermont-Gallerande left Audenarde, during the morning of the 29th, at the head of 37 squadrons, 2 battalions, and 2 companies of grenadiers. He camped that same day at Grammont, from where he immediately occupied the bridge at Ruysbroeck on the Senne. He crossed the river, on the 29th, and invested Brussels by the side of Fort Monterey.

The Marshal left Ghent, during the morning of the 29th, with 34 squadrons, 24 battalions, and 25 Swedish guns. He passed through Alost where he lodged, on the 28th, with only his cavalry. The infantry and artillery stopped in Assche, where, placed under the command of Contades, it cantoned.

De Vaux left Dendermonde, on the 28th, with 4 squadrons, 3 battalions, and 9 companies of grenadiers, moving to capture the Laken Bridge, on the 29th, by 2:00 a.m., at the latest. Once it was master of the bridge, it was to notify Contades, who was then to march immediately to join his troops to those of de Vaux. The Marshal was to follow in his turn with all his cavalry and cross the canal.

Finally, during the morning of the 28th Brézé left Tournai with the artillery destined for the siege. Four battalions, 24 squadrons, and 1 company of grenadiers escorted the guns. Brézé cantoned in Ath, on the 28th, at Hal, on the 29th, and on the 30th was below Brussels where it invested the city on the side of Alost. As to Phélippes, he left Maubeuge during the morning of the 28th, with 4 battalions, 9 squadrons, and 1 company of grenadiers. He was joined, during his march, by de la Serre, who left Bavay the same day, with 5 battalions, 6 companies of grenadiers, and 12 squadrons. The two corps, united, camped at Binch, during the evening of the 28th. They remained there during all the siege in order to mask Mons and Namur.[638]

These divisions did not contain a higher manpower than what had been previously indicated. The Marshal had directed that only

[638] Project of the Maréchal de Saxe for the investment of Brussels (de Vaux, Piece No. 64). – The dates fixed for the departure of the different divisions, according to this document, are those of 27, 28, and 29 December 1745. It is necessary to say that this project had been established to be executed on the dates that it bore and that it had been for a long time in the hand of those concerned. It is known, in effect, that in the play given to d'Argenson, on 5 December, the Marshal did not consider the capture of Saint-Ghislain as the beginning of the expedition against Brussels, the two operations were to be simultaneous. The attack on Brussels had been delayed. The circumstances that one found as a result of beginning the movement a month later had not been foreseen. For this reason, there was nothing to change except the dates on the first document. It sufficed to change the word "December" with the word "January."

vigorous and healthy men be brought along. The squadrons were reduced to 100 masters and the battalions to 450 fusiliers.[639] The troops
left with 4 days' bread, 2 days of forage, and 15 days' pay.[640] The
soldiers carried everything in their haversacks and there was no camping equipment train. The Marshal had, in effect, conceived the bold
project of cantoning his troops in the suburbs of Brussels. This had
the advantage of not encamping his troops in the open field during
this period of thaw and rain. It was, however, an excellent project on
the condition that the Allies did not make untenable the suburbs that
de Saxe proposed to occupy. If the governor burned only part of the
suburbs he could impose great difficulties on the French. In the hope
of avoiding such an unfortunate step, the Marshal wrote to de Lanoy,
Governor of Brussels, a straightforward letter in which he announced
that he was "going to execute some maneuvers with his troops near
the city of Brussels." The Marshal insinuated that it would be wrong
for de Lannoy to burn the suburbs of Brussels, according to the established practices, when the approach of French troops seemed to justify
this terrible measure. "The destruction of the suburbs of Ypres, Tournai, and Ath had not made the capture of those cities more difficult
and it is wrong to believe that the buildings outside of a fortress can
be of some advantage to the besiegers," added Maurice. Otherwise,
during the course of the 1744 campaign, had not the Marshal formally
recommended that the suburbs of Lille not be burned when threatened
by the Allies? "I have believed," he concluded, "that Your Excellency
will not disapprove the liberty that I take in writing him to conserve
such a beautiful ornament of the City of Brussels."[641] It is impossible
to determine exactly at which moment de Lannoy received this letter.
It appears that this letter arrived only shortly before the first news of
the approach of the French. However, it may be, the Allies did not
burn the precious suburbs.

---

[639] Project of the Maréchal de Saxe for the investment of Brussels (*loc. cit.*). – The real
forces of the divisions were 2,500 men (d'Armentières), 5,800 men (Brézé), 5,000
men (Clermont-Gallerande), 12,400 (Maréchal de Saxe), 1,800 men (de Vaux), 6,500
men (la Serre at Phélippes).

[640] Séchelles to d'Argenson, Ghent, 12 February (Corresp. A.F.).
– On 15 February one distributed to the troops the rest of their
pay for the month [Séchelles to d'Argenson, Ghent, 14 Feburary
(Corresp. A.F.)].

[641] De Vault, Piece No. 60.

Henry Pichat

On the fixed dates, the six divisions marched out. Varying circumstances prevented the executing of the program imposed on each of them.

To avoid indiscretions, the gates to Ghent had been closed at 4:00 p.m., on the 27th. The next day the entrance into the city was open. However, no one left except for the troops and "those that were known to belong to them."[642] The Marshal left during the morning of the 28th, in beautiful weather, at the head of his divisions where his own uhlans formed the advance guard.[643] Leaving from Alost, these cavaliers encountered the daily Allied hussar patrol. They showed them the door much more severely than ordinarily and pushed them to the gates of Brussels, itself, then they took up a post at the Afflighem Abby, watching the highway. Nothing at all of this could produce any suspicions with the Allies. The renewal of frequent skirmishes, such as this, had caused the Allies to become accustomed to seeing the French uhlans in this region. During this time, the infantry and the artillery had marched to Assche. The cavalry arrived in Alost at 4:00 p.m. They cantoned in the immediate vicinity and the Marshal took up quarters in the city. The 12 companies of grenadiers pushed to Afflighem where they took up positions with the uhlans.

During the night, the French waited in vain the arrival of a messenger from de Vaux. As of 5:00 a.m., on the 29th, no news had arrived, so Contades and his troops departed from Assche. On his side, the Marshal, with his cavalry, had also departed Alost early in the morning bringing to it the troops posted at Afflighem. This last column did not delay in joining that of Contades, which it began following. The Marshal, escorted only by a few horsemen, left before it to seek news of de Vaux at the Laken Bridge.[644] He did not find them at this last point. He then descended the left bank of the canal, stopping a few minutes at the Dillighem Abbey, then continued on. Towards 1:00 p.m., he finally encountered de Vaux at the Trois-Fontaines Redoubt, on the Vilvorde Canal.

De Vaux had marched out at the indicated hour, but misled by his guides during the night, he found himself at daybreak before the Trois-Fontaines Redobut. He immediately captured this small work

---

[642] D'Espagnac, Vol. II, p. 133 (*loc. cit.*).

[643] Crancé to d'Argenson, Alost, 28 January and on the banks of the Vilvorde Canal, 29 January at 4:00 p.m. (Corresp. A.F.).

[644]Maurice de Saxe to d'Argenson, headquarters at Schaerbeck, 30 January, 5:00 p.m. (Corresp. A.F.).

and by virtue of the means he found there, he crossed his troops to the right bank of the canal and as well as the Senne.

De Vaux had executed only part of his mission. However, he agreed to cross the canal and the Senne. To reach that point, the Marshal decided to employ the pontoons that he had brought forward with the Swedish artillery. Contades, informed at this time, turned his column's head to the left, at the level of Strombeke, to leave the Brussels Highway and move to the right on Mariensart. At 3:00 p.m., Contades' infantry began to cross the canal in the same manner of de Vaux's infantry. The cavalry then arrived, but the artillery and the precious pontoons did not arrive until 4:00 p.m., because they were delayed by the bad roads that they had had to follow once they left the highway.[645] Under the protection of de Vaux's troops arrayed in battle formation facing Vilvorde, the French worked to throw a bridge over the Senne. As night fell, their work was not yet completed, however, despite this, four battalions had already crossed to the right bank. The troops bivouacked in the position where the night caught them.

If he had finished in encountering de Vaux, the Marshal still awaited de Clermont-Gallerande. However, it was known that this latter was to find himself, during the 29th, in the region northeast of Brussels. This absence disturbed the Marshal, because, for several minutes, he had thought he had seen the arrival of de Clermont-Gallerande in the vicinity of the Allies, who had suddenly appeared. In Brussels, the Allies had learned, at 8:00 a.m., on the 29th, of the approach of the French troops from the Dilighem monks. General van der Duyn, commander of the Dutch troops in Brussels, had thought that the moment had come to execute the orders prescribed by von Waldeck in his conference of 16 January. He therefore moved, with 8 battalions and 2 cavalry regiments, to the Vilvorde Canal, as he had been ordered to do.[646] When he arrived in the Evère plain, the French troops were spotted. De Clermont-Gallerande was expected to arrive from that direction, but it was these Allied scouts who came within

---

[645] Ibid. – De Vault, p. 140 and following. – The Marshal adds in the previous letter, "I shall have to take three battalions to Vilvorde with some ceremony and the Lord Ferret wished to capitulate at Grimbergen with the honors of war, which I did not wish to accord him. Thus I left him there to move on to my principal objective, which is Brussels." – During the night of 1-2 February, two of the Vilvorde battalions left this small fortress and Ferret evacuated Grimbergen. He was not accustomed to granting the honors of war to partisans, like Ferret, whose troops were recruited, for the greatest part, from among deserters.
[646] von Waldeck, 29-30 January.

pistol shot distance. The French officers forbad firing on them, however, because they thought these troops were the Beausobre Hussars, who marched in the vanguard of the expected division. The French and Dutch confusion was quickly dissipated. "[Van der Duyn]," wrote Waldeck, "sent an officer to see if the Allied troops were in their posts, and finding almost all the [Allied] troops had marched directly on Antwerp, returned his soldiers to Brussels."[647]

The Marshal thought that the approach of Clermont-Gallerande provoked this retreat, to which, one will see, von Waldeck assigned other causes.[648] At this time, the Prince complained bitterly that the positions ordered, on the 16th, had not been occupied. "This maneuver, if executed, would probably have saved Brussels," he said.[649] These were superfluous regrets. Nothing proves that if "the maneuver" had been executable and that the departure of the Hessians and Hanoverians had not rendered it impossible, it would have sufficed to assure the safety of Brussels. The Marshal had established his plan considering the withdrawal of the Hessians and Hanoverians; but if these latter troops had not moved back to Antwerp, Maurice would have adopted other dispositions, because he was firmly decided to attempt this enterprise.

At 10:00 a.m., on the 30th, and because of work completed during the night, the Mariensart Bridge was completed.[650] Four battalions and 12 companies of grenadiers crossed immediately and occupied Haerer, facing Trois-Trous, on the right bank of the Senne. The rest of the infantry and Contades remained on the left of the canal. All the cavalry crossed the bridge and moved to unite with the troops of de Vaux on the Evère Plain so as to march on Brussels. The French line had barely formed when on its left the heads of Clermont-Gallerande's columns finally appeared. This general officer had cantoned at Grammont. Conforming with the Marshal's orders, he had immediately sent a reconnaissance on Ruysbroeck. Local peasants had informed him that it was impossible to cross the Senne at this small village where the bridge was in a very bad state. Clermont-Gallerande could only cross the river at Hal, and this had taken him a long way from his line of march, producing his delay.

---

[647] *Ibid.*

[648] Maurice de Saxe to d'Argenson, Headquarters at Schaerbeck, 30 January, 5:00 p.m., (Corresp. A.F.).

[649] Waldeck, 29 and 30 January.

[650] Maurice de Saxe to d'Argenson, Headquarters at Schaerbeck, 30 January, 5:00 p.m., (Corresp. A.F.).

A few hours later d'Armentières arrived. He had also completed his mission. It is impossible to know how aggressively he had attacked Nivelles, an easy victory, which he did not capture.[651]

Finally, the uhlans, the Grassins, and the hussars searched Louvain and Malines. They found, as one had expected, that these cities were completely evacuated by the Allies.[652]

The junction of the Marshal with his two lieutenants completed the investment of Brussels on the right bank of the Senne and on the north of the city. The French sought to move as close as possible to the precious suburbs that were still intact. The Marshal decided to seize the Laken Bridge. The French could not use the passages at Mariensart to definitively assure communications between the two banks of the canal and the Senne. The bridges were much too far from the highway from Ghent to Brussels, the only road usable by the re-supply wagons. Contades and the infantry left on the left bank lifted the pontoons at Mariensart and then moved to Laken. To execute this movement by the shortest route, Contades had to pass under the fire of the Trois-Trous Redoubt. He summoned it to surrender. The commanding officer, with 150 men surrendered. At midnight, the French troops occupied the Laken Bridge, after having chased away a small Allied post. Contades immediately threw over the Senne, before Laken, the bridges brought from Mariensart. This night the soldiers bivouacked in the open. They showed themselves to be very vigilant, because a cavalry brigade alone covered the Marshal's headquarters, provisionally installed at Schaerbeck while he waited for it to be possible to move on Laken. However, the Allies did not move.

Finally, the 31st, the troops entered their cantonments. The suburbs had remained unharmed. The soldiers installed themselves

---

[651] Von Waldeck wrote in his journal on 4 February: "The garrison of Nivelles, after having sustained an attack by a corps of 5,000 to 6,000 men, which had attacked this city with artillery, and had bravely defended itself for some hours, drove back some attacks, and after the enemy's departure, withdrew to Namur." — The Prince reproduced, on 14 February, *in extenso*, a letter addressed by the engineer Villarmont to d'Armentières. This document, intercepted by the Allies during the siege of Brussels, relates a different story. One reads in it: "Our siege was very advanced for what concerned us. But the artillery, as usual, was very slow. It was not until 11 p.m., that two mortars were in battery and they did not produce a great effect. Most of the bombs had burst upon leaving the mortars, as before Nivelles, as a result of old fuses." According to d'Espagnac, Armentières passed before Nivelles, but he did not attack this small fortress out of fear of delaying his march on Brussels [d'Espagnac, Vol. II, p. 132, Note 1 (*loc.cit.*)].

[652] von Waldeck, 4 February.

Henry Pichat

strongly after two days of difficult marches and two nights spent in bivouac. A persistent rain had already begun to fall.[653] Everywhere the French entrenched. The fronts of the cantonments looking on the fortress were put in a state of defense and the different quarters linked themselves with communication trenches. The cavalry occupied the villages closest to Brussels and the infantry occupied the suburbs proper. During the evening of the 31st the four divisions completed their installation, a process that the Allies did not attempt to disrupt.[654]

However, de Brézé did not arrive at the rendezvous. During the evening of 1 February, accompanied only by Gourdon, he finally presented himself at the Marshal's headquarters. The troops of this division and the artillery equipment destined for the siege were, at this time, stopped in Anderlecht.[655] The bad state of the roads and the abominable weather had slowed the march of this division. On 2 February, Brézé's soldiers led by de la Suze finally occupied the cantonments that had been reserved for them. As for the material, it moved into its park at Laken with great difficulty. One soon saw that it could not be left there without it becoming deeply stuck in the mud. Commissioner Labinon made the necessary dispositions to place it along the highway, near the canal, before Laken where it could remain in the vicinity of the fortress. This work was only completed with difficulty by the 7th. During the process, a large number of traces and prolongs were broken. In addition, a number of guns were toppled, but this occurred without much damage to the equipment.[656]

On 2 February, completely entrenched in their cantonments, the troops began making fascines and the siege corps established its definitive organization. The first bread convoy had arrived from Ghent, followed by the mobile hospital, which was provisionally installed in Dilighem Abby. Crancé made all his dispositions in order to begin regular distributions to only five divisions. That of Phélippes

[653] Crancé to d'Argenson, Laken, 31 January, 8:00 a.m., and 1 February (Corresp. A.F.).

[654] D'Espagnac recounts, however, that the fortress has begun to cannonade the French infantry: "The Marshal immediately informed de Lannoy that he would send him as many heated shot as Lannoy fired on him. This threat imposed on the fortress and the French infantry remained peacefully in its lodgings." [d'Espagnac, Vol. II, p. 140 (*loc.cit.*)]. – No document confirms this allegation. It appears only that some parties of hussars pushed a reconnaissance on Laken and they were driven away with musket shots.

[655] Crancé to d'Argenson, Laken, 1 February (Corresp. A.F.).

[656] Maurice de Saxe to d'Argenson, Laken Castle, 5 February, 5:00 p.m., and Laken headquarters, 7 Feburary (Corresp. A.F.).

was to be supplied, in the Austrian Hainaut, by its own means. From the departure, the troops drew their forage from the regions that they had successively occupied. By the orders of the Marshal the French began to send requisitions into the regions around Brussels, Louvain, and Malines to form the first magazine of 200,000 rations at Laken, a second of 30,000 at Asche and a third at Alost. This latter was, above all, destined to furnish the troops in the rear area. The French also imposed a meat requisition on the region such that each battalion and each cavalry regiment of four squadrons would receive 200 livres of meat a day. The cattle were to be delivered on foot and the distributions assured by a commissioner of war.[657]

On the line of investment, the troops found themselves dispersed as follows: At Laken and the bridge – the headquarters, the artillery park, the Vernon Auxiliary Militia companies, the 3rd Battalions of the Paris, Montargis, and Saint-Maixent, and Nantes Regiments, plus 5 battalions and 10 squadrons. The mobile hospital was provisionally located at Dilighem with a guard of 50 men. In the Flanders suburb d'Armentières commanded 8 battalions and 8 squadrons and Beauffremont had 4 battalions and 18 squadrons. In those of Namur, Louvain, Ixelles, Etterbeck, Saint-Joorist-en-Hode there were 12 battalions and 25 squadrons placed under the orders of Clermont-Gallerande. Ten squadrons of Carabiniers taken from d'Armentières' command occupied a number of posts at Trevuern, Ophem, and the vicinity. The 12 battalions and 25 squadrons, under Brézé and la Suze, cantoned in the suburb of Schaerbeck. The four squadrons of Beausobre Hussars were posted at Steenockerzele and the foot and mounted Grassins were in Louvain. Finally,500 fusiliers occupied Trois-Fontaines. The Marshal commanded the army assembled before Brussels. Brézé directed the siege operations, assisted by Labinon and Gourdon as Directors of the Artillery and Engineers. Crancé, Ordnance Commissary, filled the functions of Intendant of the Army. As for Séchelles, Indendant of Flanders, he assured the supplies from Lille and Ghent. For the moment, he prepared the reinforcements of artillery and ammunition in order to send them forward upon the Marshal's requests.[658]

According to Allied deserters, the Governor of Brussels had contented himself with prematurely fatiguing his garrison by having it take up arms daily at 4:00 a.m. and 8:00 p.m., to put a battalion at each

---

[657] Crancé to d'Argenson, headquarters at Laken, 1 and 2 February (*Ibid.*).

[658] De Saint-Périer had the order, from 23 January, to prepare at Douai, a convoy of siege equipment of the same composition that de Brézé had organized in Tournai [Maurice de Saxe to d'Argenson, Ghent, 23 January (Corresp. A.F.)] and Séchelles to d'Argenson, Lille, 2 February (*Ibid.*). – The division of troops along the line of investment was addressed, on 2 February, to d'Argenson by Graulet.

Henry Pichat

gate and to line the ramparts. Von Waldeck, on his side, deployed, although too late, with great activity. On the 29th, van der Duyn, Commander of Dutch Forces, pressed to keep the Prince current on the events that occurred below Brussels. The Prince received news from La Haye, on the 31st. He reached Antwerp, moving at all haste, but contented himself in addressing to the absent von Waldeck the letter from van der Duyn. On 1 February, at noon, Waldeck visited Lord Dunmore, Governor of Antwerp. This latter had also just received bad news. They removed from the Prince the last hope that he could successfully oppose de Saxe. There, also, Waldeck learned of the complete abandoning of Louvain and Malines by the Hessians and the Hanoverians assembled, at that time, in the vicinity of Antwerp, while they awaited their pending embarkation for England. "Informed of everything," he said, "I asked the generals that were in Antwerp to come to me, in order that we might confer. Being the senior general, I proposed, if it was not possible to assemble all our troops and to attack the enemy before he was reinforced and established in his quarters. Dunmore, Mölck, Prince Ferdinand von Hesse, and Crawford were of this opinion; however, the first added that he had positive orders to move to England the English and Hessian infantry, that he could not delay the execution of this order by providing troops for this operation. However, General Ilten represented that with so few troops it was not possible to undertake anything, having left his regimental artillery at Herenthals and that of the English was already embarked. I did not have a single cannon at my disposal; also, I thought I would overexpose the forces of the State if I undertook to march forward with them alone. I was reduced to taking, the next day, a post at Malines, the only debouch which remained for us to move against the enemy, be it to make a favorable diversion, be it to assist the garrison if it executed a sortie. Finally, I obtained from General Ilten the concurrence with his troops to form a detachment of 1,000 infantry and 200 horse which I proposed to send to occupy this post. Upon leaving his conference I received word that some irregular enemy troops had entered Malines, so it was decided that the detachment destined to go there, would march to the Walhem Bridge and there it would send a reconnaissance if the enemy still occupied it and in what manner it was posted there, to see if one could force it."[659]

---

[659] Waldeck, 1 February. – The Dutch siege artillery had been sent to Dorth around 17 October 1745 [Koenigsegg to Their High Powers, Beaulieu, 17 October (Corresp. G.H.)]. As for the Republic's field equipment, it was almost entirely in Brussels,

Maurice De Saxe's 1745 Campaign in Belgium

This picture which we have intentionally presented in its entirely shows the complete disarray in the middle of which the French attack had surprised the Allies. It established that Maréchal de Saxe had chosen the moment to attack with perfect clairvoyance. Finally, it emphasizes the effectiveness and the success with which the measures taken in Ghent had concealed the French preparations from the Allies.

Von Waldeck does not tell us if the reservations made by Dunmore and the Hanoverian General Ilten convinced him of their ill-will. However, they would soon give von Waldeck new proof of their ill-will. Be that as it may, the measures taken by Waldeck left one to think that this conference had confirmed him in the resolution to not count, in the future, on more than his own forces. The Prince sent to Breda and Ruremonde the order to send five regiments on Antwerp. "I have also," he wrote, "sent notice to all the regiments to incessantly bring in recruits, as well as the soldiers on leave to return and proposed to the officers to assemble at Antwerp all those who belonged to the garrisons of Mons, Charleroi, and Brussels. I would form them into two battalions, to which I will give weapons that I will draw from Breda, as well as the munitions of war that we do not have."

At 6:00 a.m., on the 2nd, he occupied Walhem with the detachment formed the day before.[660] Learning that Beausobre had only passed through Malines without stopping, von Waldeck immediately posted 1,200 men in this city and reinforced them with two battalions drawn from Vilvorde. These troops were to camp at the city's gates, facing Brussels, hold the bridges at Eppeghem and Semps, and spare nothing to keep von Waldeck informed on the French activity in that region. In case of attack, and only in the last extremity, they were to withdraw on Walhem. Upon his return to Antwerp, the Prince found comforting news carried by a courier from General Baronyay. He brought the first reinforcement of Imperial troops and sent to von Waldeck letters destined to von Kaunitz, Governor General of the Lowlands. "I opened them," reported the Prince, "and learned from them that this corps had stopped in the land of Cologne, because of the lack of money to continue. After having conferred with General Mölck, I borrowed from van der Howe, Burgomeister of Antwerp, the sum of 100,000 florins, which I sent by some officers to this general, with the most pressing orders to accelerate his march as much as possible. As there were no magazines prepared on the new route that this corps had

which the Marshal knew.
[660] Waldeck, 2 February.

to follow, as a result of the investment of Brussels, we brought together a sum, General Mölck and I, which was given to some businessmen who were ordered to organize these magazines." If one considers the particular situation of the Dutch in the Coalition, if one remarks that the garrison of Brussels contained only a few Hessian or Hanoverian troops, but was almost totally troops from the Dutch Republic, one finds a most singular contrast between the zeal of von Waldeck and the indifference of Ilten. In sum, the bad habits already noted in the conferences of 1745 were seen again now. As previously, they could only lead to the most unfortunate results.

The Marshal had chosen as the point of attack the part of the rampart covered by the Brussels horn work, because the ground was dry and sandy which would permit the siege work, and "where one could easily move up the guns."[661] This was also one of the strongest sectors of the defense. Finally, this horn work overlooked a sort of small plateau. After having carried it, the French established their artillery there so it could command all the fortress. Unfortunately, the persistent rain and the thaw flooded all this area. Masters of the canal as far as Vilvorde, the French could open some of the sluices and lower part of the flooding. The French could only manage, however, to definitively dry the communications from Laken to Schaerbeck only by operating the sluices contained in Vilvorde Castle. While waiting, to move the cannon of the park over the ground of the attacks, it was necessary to move down the left bank of the canal as far as Vilvorde, cross the bridges in this village, and then move up the right bank; in total, a march of about four miles, half of which was over bad roads. It was important, therefore, to capture Vilvorde Castle as soon as possible. However, at the time this post, though very solid, lacked supplies and possessed only a very weak garrison.

On 4 February, the Saxe-Volunteers and the French post at Trois-Fontaines presented themselves before the castle. "We threw several bombs with our howitzer into it and the commandant said he would surrender if an aide-de-camp from von Waldeck did not arrive, apparently, bringing to this officer orders to hold out," said the Marshal.[662] Otherwise Maurice did not explain how this emissary could

[661]La Graulet to d'Argenson, Laken Camp, 6 February, 1:30 p.m. (Corresp. A.F.).

[662] Maurice de Saxe to d'Argenson, Laken Castle, 5 February, 5:00 p.m. (Corresp. A.F.). – The "howitzers" of which the Marshal spoke are the howitzers that served for the first time before Brussels which one had captured from the Allies at Fontenoy. The French siege material contained no howitzers and none were found in the siege equipments of its armies.

pass, since from Malines to Vilvorde the dragoons and hussars swept the entire region. The truth is that the old Cornabé had left Walhem with a reconnaissance of 100 horse furnished by the detachment placed there by von Waldeck and of which Maurice was still unaware. Pushing as far as Vilvorde, the adjudant of the Prince had found there a "French drum major sent to summon Vilvorde to surrender. The commander responded that he had orders to defend himself."[663] Cornabé had acted to draw the Marshal's attention away from the north side. Otherwise, the Allies prepared to augment the troops initially thrown into Malines. These projects presaged the reinforcement of Vilvorde. The Marshal, placed on his guard, pushed for the capture of the castle. Contades recommended an attack that same day. It was not until the 6th, and before the arrival of the cannon, that the work capitulated.[664] It was immediately occupied. The French then operated the sluices of the canal and one did not delay to note that in spite of snow, rain, and hail falling endlessly the water started to drain away. The ground selected for the attacks and the communications with Laken were soon entirely dry.

To remedy the loss of Vilvorde, von Waldeck reinforced his post at Semps. He apparently decided to extend to the side of Louvain. One hundred horse, 50 dragoons, and 100 infantry already held the bridges at Werchter and Keerberghen, and daily patrolled along the Dyle. The Hessian cavalry was cantoned on the Rupel, covering all the region from Malines to the Escaut. The Hanoverian cavalry extended along the Neth from Lierre to Antwerp. These forces sufficed to assure the security of these various regions. Von Waldeck conserved, therefore, complete latitude to extend towards Louvain. He thought about it, but he began by executing some defensive works at Malines and also occupied the debouches from Pasburgh and Bonheyden to hold the French reconnaissances at a distance, when they were attracted by the Allied reoccupation of Malines. In sum, von Waldeck sought to retake Louvain. His plan did not gain the assent of the Allied generals. "I believe," he wrote, "that we can undertake something on this fortress. I proposed it, but for the same reasons stated before, one was of the attitude that we were not in the state to undertake anything before the arrival of the Imperial troops." Desiring to execute his project, the Prince sent to the Meuse to hasten the movement of the anticipated

---

[663] Waldeck, 4 February.

[664] La Graulet, Lake, noon on 5 February to 6 February, at 1:30 p.m. (Corresp. A.F.).

reinforcements.[665]

This general pandemonium did not escape the Marshal. He sent d'Armentières and a brigade of cavalry to reinforce the Grassins at Louvin. It was important that the Allies not occupy this city, because the siege corps drew most of their subsistence from this region. At the same time, on 6 February, Phélippes, still peacefully encamped at Binch, received orders to send de Relingue to Genappe with the major part of the division. De Relignue drew 5 companies of grenadiers and 4 battalions, which were going to pass to the siege corps, camping at Dieghem and Haeren. Phélippes returned to Maubeuge with one battalion of infantry and a regiment of dragoons. He thus found himself in a position to watch the French frontier which was more threatened by the garrisons of Mons and Charleroi than was the siege corps. Otherwise, everyone closed in on Brussels, while Relingue, camped at Genappe, could continue to watch both Mons and Charleroi. These measures also removed the difficulty caused by the suffering of the 6,000 men of Phélippes who drew their subsistence from Austrian Hainaut.[666]

During the night of 7/8 February, the French constructed, without interference, the first parallel about 160 toises from the covered road of the horn work. The work of the siege had begun. At the price of considerable efforts, Labinon had put in line the first convoy of 12 cannon and 6 mortars, with their munitions, along the road running along the left bank of the canal. The roads having been repaired, he hoped that the batteries would be completed on the 9th, such that the materials could be brought forward on the 10th and fire opened on the 11th. A second convoy of 12 24pdrs with 6 mortars would arrive later because one was expecting, in Ghent, a resupply of powder and munitions coming from Lille and Douai. Two hundred wagons were, otherwise, ready to bring everything forward to Brussels.

Finally, during these last days, the siege corps had completed its organization. The French installed, at Schaerbeck, a trench ambu-

[665] Waldeck, 4, 5, 6, & 7 February.
[666] Maurice de Saxe to d'Argenson, headquarters at Laken, 7 February (Corresp. A.F.).
– On the subject of the difficulties of provisions, Séchelles had already presented some observation to the Marshal. "He had responded to me," wrote Séchelles, "that he had sent instructions to de Relingue to have him furnish bread in the land and that Bernier, who had the detail of this corps, would compensate for the functions of Commissary Saint-Marie, who had returned to Maubeuge with de Phélippes. I have taken the liberty of telling him that my confidence in Saint-Marie, who he knows better than I, is not great; but unfortunately I have no commissary that I can send there." [Séchelles to d'Argenson, Camp at Laken, 8 February (Corresp. A.F.)].

lance and the mobile hospital moved to Vilvorde. However, there were few sick, despite the terrible weather.[667] The infantry and cavalry found themselves under cover in their cantonment, food was abundant, and food distributions were regular. At the request of the Marshal, Séchelles sent to Lille a considerable number of shoes. The King had authorized that the first distribution be made free.

The various movements of material, fascines, food, and forage necessitated a large number of horses because of the bad weather. "You will not be shocked," wrote Séchelles, "when I tell you that it requires six strong horses to draw three sacks of oats in a wagon when one leaves the road."[668] One had been able to shelter the horses of the cavalry and the artillery reserve. It had been impossible, because of a lack of places, to extend this measure to the 6,000 horses requisitioned for the needs of a very loaded service.[669] The French lost many animals because of the bad weather.

Finally, on the 9th, the weather improved and the French were 60 toises from the covered road. The following day, at 2:00 p.m., the first battery of mortars began fire when the Marshal received a peace offering from von Kauntiz. "Without entering into the superfluous reasoning," said the latter, "I am ready to capitulate if the garrison can leave with the honors of war and that the other points of the capitulation can be based on the same principal."[670] So many preparations, fatigues, and efforts could not be paid by such an inadequate result at the time when the siege was going to enter the active phase. "You made the error of putting a garrison in Brussels which is not a defensible fortress. There no longer exists any means of helping Brussels. It is just that we benefit from all these advantages," replied the Marshall in a letter that was as firm as it was skillful.[671] It was intended to intimidate von Kaunitz while giving him assurances that the city ran no

[667] "I cannot express the horror of the weather that we had experienced for 36 hours. The rain, the snow, the hail, everything was unleashed. The vicinity of the camp is impracticable. I spent 6 hours on a horse as I visited all the establishments." [Séchelles to d'Argenson, Camp at Laken, 8 February (Corresp. A.F.)].

668 Séchelles to d'Argenson, Camp at Laken, 8 February, (Corresp. A.F.).

[669] *Ibid.*

[670] A copy of this letter, dated Brussels, 10 February, was sent by Maurice to the Minister on the following day. [Maurice de Saxe to d'Argenson, headquarters at Laken, 11 February (Corresp. A.F.)]. – He attached to it a response which he had immediately sent to von Kaunitz. The latter is dated from the headquarters at Laken, 11 February. [671] De Broglie has published a document, with a few gaps, in *Maurice de Saxe et le marquis d'Argenson* (Vol. I, p. 42).

Henry Pichat

risk of being devastated on the condition that the governor decided to quickly capitulate. He had wished, above all, as he said, "to furnish a clear statement that would serve as a justification to those that he commanded."[672]

It appears that the Marshal's response communicated a new ardor to the defenders of Brussels. From the 11th, the fortress's fire became more murderous than it had been up to that point. The garrison undertook a bold effort to demolish the siege works. Contrary to expectations, to this point only the mortars were in action. No cannon had yet opened fire even though the heads of the saps were within a dozen toises of the covered road of the horn work. During the night of the 11th/12th, 2,000 men of the garrison launched a sortie.[673] In order to prevent any others, it was resolved to make untenable the covered way, where the garrison could assemble forces for other sorites. The Marshal had raised two trench cavaliers[674] from where the infantry's fire would plunge into the road. He also ordered the construction of two enfilading batteries that could at the same time open a breach in the horn work proper. Otherwise, it was important to quickly capture the covered road, because it ran along a crest that masked the horn work whose exact force and disposition were unknown to the French. "Up to the present," said the Marshal, "neither de Brézé, nor Gourdon, nor Löwendahl, nor anyone else could say what lay behind this [covered] way. It is marked as a horn work on the plan, but it is not one. It is, I believe, a plateau, which will be advantageous for us as soon as we are masters of the covered way."[675]

On the 13th, only, grenadiers and dragoons definitively occupied the cavaliers that the projectiles and the garrison's workers had earlier succeeded in knocking down.[676] At the same time, four 12pdrs opened fire and cleared the Allies from the covered way.[677] The following day the French soldiers established a lodging in the covered

---

[672] Maurice de Saxe to d'Argenson, headquarters at Laken, 11 February (Corresp. A.F.).

[673] Reports of Brézé and La Graulet to d'Argenson, 9, 10, 11, and 12 February (Corresp. A.F.).

[674] Translator: A cavalier is a work raised 10-12 feet higher than the rest of the work. The intent of these cavaliers was to give the French an elevated position from which they could fire into the covered way.

[675] Maurice de Saxe to d'Argenson, headquarters at Laken, 12 February (Corresp. A.F.).

[676] Report of Brézé and La Graulet, 11, 12, and 13 February (Corresp. A.F.).

[677] *Ibid.*

way. They pushed forward some sapper's hooks [used to position fascines]. This made it possible for the French to determine that they were faced by a horn work, a serious obstacle, half reveted with a demi-lune.

These last works were very slowly carried out. The French troops, however, showed very good morale. If the Marshal was pleased to see the soldiers "always lusty and content," he complained of the engineering and artillery officers. "If there were more intelligence among them," he said, "our operation would have been better executed. It should be hoped that all that I say to these and the others higher up will have its effect."[678]

The indefatigable activity of von Waldeck made these delays regrettable. From 7 February, the two battalions formed with recruits and the men recalled from leave served in Antwerp and the forts of the Escaut. On the 8th, the Prince of Hesse embarked for England with his infantry; but on the 10th, the garrison of Nivelles entered Malines along with a regiments of hussars that escaped from Brussels before it was invested. New posts lined the Dyle between Malines and Aerschot. The Dutch Dorth, Schlippenbach, and Hope Infantry Regiments moved closer to Walhem, Duffel, and Herenthals. Finally, the head of the Imperial column had reached Ruremonde, where, in truth, it suffered great difficulties in crossing the Meuse as a result of the ice.[679]

The Marshal was not unaware of these movements. He doubted, however, that von Waldeck would attempt a battle to break the Brussels blockade. The ground occupied by the French permitted them to effectively utilize their numerous cavalry, which to that point, had suffered from neither the bad weather nor from the siege. In case of an attack von Waldeck could not conceal any of his movements, at least on the side of Malines. As a result, Maurice resolved to leave the Prince only the debouch from this village in case, despite expectations, he attempted the attack that he seemed to be preparing. The Marshal moved Relingue's troops into Louvain. Löwendahl took command of the five battalions and the 32 squadrons that were thus assembled.[680] The Grassins and the Beausobre Hussars were also placed under his orders. Finally, Vilvorde was reinforced with two battalions because

---

[678] Maurice de Saxe to d'Argenson, headquarters at Laken, 14 February, 4:00 p.m. (Corresp. A.F.). – "The troops are very content." [Crancé to d'Argenson, headquarters, 13 February (Corresp. A.F.)].

[679] Waldeck, 10, 11 & 12 February.

[680] Löwendahl to d'Argenson, Laken, 15 February (Corresp. A.F.).

of the importance of the strength of the Allies in Malines. The Marshal decided, then, that if the Allies obliged Löwendahl to evacuate Louvain, one would maneuver as necessary. The closest two battalions of troops would immediately reinforce Löwendahl's infantry, who would take up a position along the Woluwe stream, from Dieghem to the Louvain Highway. At the same time, the Grassins would occupy Saint-Stephens-Woluwe to block the crossing of the stream. Finally, 80 squadrons of cavalry would form a second line behind all this infantry. The troops reconnoitered their emplacements on the line of battle that were thus chosen. They also examined the roads destined to take them there.[681] This disposition presented the advantage that if von Waldeck attacked the French would not have to move any infantry or empty any trench, nor the posts covering the cantonments. The garrison of Brussels could not, therefore, combine its efforts with those of the Prince to attempt a diversion on the French rear.

It appears, however, that these preparations came at exactly the right moment. Von Waldeck resumed his initial projects on Louvain, which appeared to him to be the heart of the situation. Before the ill-will of the Allies, the Prince was of the view that he was obliged to momentarily delay the execution of his plan. On 13 February, he went to Malines to confer anew with General Zastrow, the commander of that fortress. There, von Waldeck learned of the reinforcement of Louvain and the arrival of Löwendahl in that city. He could no longer, henceforth, doubt the intentions of the Marshal and in these conditions he was obliged to await the arrival of Imperial reinforcements. Then von Waldeck resolved, for the moment, to attack the French logistics. He sent a free company into Aerschot. He sent a courier to the Caroly Hussar Regiment, which had been sent forward by Baronyay as he crossed the Meuse at Maëstricht. These horsemen received the order to send large parties to Genappe and Louvain, from where they could reach Saint-Trond and capture French convoys.[682]

Von Waldeck was well inspired to have taken such a fortunate resolution. If the Caroly Hussars could complete their mission, they would absolutely make the daily situation more difficult. Despite all possible activity, one could only resupply the troops with difficulty

---

[681] De Vault, p. 158. – The troops designated in the order of battle prepared for the circumstance formed a total of 22 battalions and 89 squadrons (de Vault, piece no. 74). – D'Espagnac sent a complete plan of what the Marshal had ordered [d'Espagnac to d'Argenson, headquarters at Laken, 13 & 14 February (Corresp. A.F.)].

[682] Waldeck, 10, 11, 12 & 13 February.

from the three magazines established by Crancé. It was anticipated that by 2 March these magazines would be empty, as the surrounding territory was more and more exhausted. The French could then only count on the magazines in Ghent. It would be very difficult, in this case, to organize regular convoys. One could judge this as at the same moment Séchelles had been obliged to momentarily delay certain shipments, the bad weather and the movement of the artillery, having destroyed the roads.

Happily, from the 15th, the French siege operations entered the last phase. That day 18 cannon and 6 mortars bonbarded the defenses of Brussels and began to gain superiority over the defenders. The weather had also turned to a dry cold, since the 13th, and this permitted the work to advance with more rapidity and perfection.[683] On the 16th, all the batteries were completed that were destined to fire on the horn work and the main body of the fortress. The sap already descended into the ditch of this work the day after the garrison abandoned the demi-lune. This permitted the French troops to make an immediate lodging.[684] The Marshal hoped that, on the 17th, the cannon would complete a sufficient breach in the horn work to permit an assault.[685] However, it was only during the night of 19/20 February that one was able "to attempt" the work. Opinions were divided on the degree of accessibility of the two breaches opened by the artillery, whose projectiles had also caused the collapse of the parapet.[686] In sum, it was only a reconnaissance that one was about to attempt. A bit more was

[683] Crancé to d'Argenson, Headquarters, 13 February (Corresp. A.F.).
[684] Bulletins & reports of La Graulet and Bréze, 14, 15, & 16 February (Corresp. A.F.).
[685] "We currently cannonade the horn work. I believe that that it will be broken tomorrow. If it is, I shall attack it. It is necessary to take it with a vigorous attack. It is capable of a considerable defense. Perhaps it will only be a mediocre defense. And as this work is open in the rear, it is not possible that the garrison, after having been chased out of it will not return. We will advance with caution, and we will put to use all that appears to me to be proper to manage our troops." [Maurice de Saxe to d'Argenson, headquarters at Laken, 16 February (Corresp. A.F.)].
[686] Reports of La Graulet and Brézé 16, 17, & 18 February. – Our breaches were not found sufficiently practicable today to allow an assault on the horn work. This is why I have put it off to tomorrow, where I expect that they will be complete. This work is open to the rear and has a gentle slope. All the deserters say that the garrison has 2,000 men behind to support it which will make this attack a bit hot. However, we have directed all the military art against it and all the vigor of which our troops are capable and our will. We have a breaching battery firing on the body of the fortress extending the left bank of the horn work. This breach is very advanced. If we remain masters of the horn work, as I hope, I expect the city will quickly surrender." [Maurice de Saxe to d'Argenson, headquarters at Laken, 17 February (Corresp. A.F.)].

required for the French to make themselves masters of the work.[687] By virtue of the vigor and audacity of the French troops, who that night justified the prodigious praise of the Marshal, the action had a decisive result.

In the various quarters of the cantonments and over all the extent of the line of investment, one took, before the attack, all the precautions so that the garrison could not force any part of the circle that enclosed it. One designated for the assault four companies of grenadiers, 100 men in a picket, and 100 dragoons. All these troops were massed in the ditch. At 8:00 p.m., and on each of the breaches in the left and right faces of the work, a sergeant climbed up first, escorted by a few grenadiers without encountering any obstacle. On each side, two companies of grenadiers followed immediately with workers. Instead of stopping to cover the sappers charged with preparing lodgments in the breaches, the grenadiers continued to climb. Suddenly they arrived on the platform of the work crying *"Vive le roi!"* and calling forward the pickets and the dragoons, who had remained in the ditch. These latter did not wait. They climbed the breaches in an instant. Without thinking of covering the workers who set to work, they mixed with the grenadiers. The French were masters of the horn work because of the rapidity of the action had completely surprised the garrison. Already the French soldiers, carried away by their ardor, fell on the palisade of the covered way of the main body of the fortress, which ran behind the gorge of the work when the defenders arrived in force. A stubborn battle began. It lasted for more than an hour and at the end of it the French were turned and forced to evacuate their conquest. They descended the breach without the least confusion "asking that they be allowed to go forward again."[688] The French had about 100 combatants on the ground. This check was immediately followed by a resumption of violent fire by the garrison. However, this was their last effort. The French troops had barely returned to their cantonments when the garrison sounded the "chamade"[689] and raised the white flag.[690]

On 15 February, von Waldeck learned in a certain manner of the approach of Imperial reinforcements. On the next day, the confer-

---

[687] Brézé to d'Argenson, before Brussels, 19 February 10:00 p.m. (Corresp. A.F.).
[688] Séchelles to d'Argenson, Ghent, 20 February, 2:00 a.m. (Corresp. A.F., A.G.).
[689] Translator: The "chamade" was a call to parley.
[690] La Graulet to d'Argenson, camp before Brussels, 20 February (Corresp. A.F.). – A "detail on the attempt on the horn work of 19 February" is attached to this letter. – Séchelles to d'Argenson, Ghent, 20 February, 2:00 a.m. (Corresp. A.F.).

ences resumed. "I assembled with me," wrote the Prince, "the Generals Mölck, Ilten, Aylva, Hompesch, Zastrow, and Dalwig to coordinate the means to undertake something to deliver Brussels upon the arrival of the reinforcements. After having considered our position and that of the enemy, it was agreed by a majority of the voices that it was necessary to begin by seizing Louvain, which would give us the facility of going against them by the two roads from Louvain and Malines." General Ilten accepted the plan and provisionally made arrangements for the arrival of Baraonyay's corps, delaying the plans for the attacks to another conference. They saw that Maréchal de Saxe had reinforced Louvain strongly.

The Allies designated the cantonments reserved for the Imperial troops. They reconnoitered the crossing points over the Dyle and placed posts there. They "attempted an attack" on Louvain, attempting to force, in three simultaneous attacks, the Tirlemont, Diest, and Rousselberghe Gates. Löwendahl, warned, was on his guard and the Allied attacks failed.

On the 17th, three Imperial battalions with General Lilliers entered Westerloo, followed the next day by two others and a regiment of dragoons. Immediately after the arrival of this first column, the Allies prepared a new attack on Louvain. "As soon as they entered," said von Waldeck, "one made new provisions to deliver Brussels." Nothing was realized from these bellicose intentions, as General Ilten persisted in showing the most insurmountable indifference.[691]

Confident in the energy and capacity of van der Duyn, as well as the valor and patriotism of the Brussels garrison, most of which was Dutch, von Waldeck hoped, despite the new delay imposed by Ilten's ill-will, to arrive in time to break the siege of the city. "I wrote to van der Duyn," he said, "by a peasant to whom I promised great rewards if he could enter Brussels; that I hoped to be with him on the 20th, awaiting the succor towards that time, which could not arrive because of the ice. I received a response, through the same peasant, a letter in which he informed me that he was going to raise the white flag the same day, but that he would do everything to drag out the capitulation

[691]"After the project for the attack on Louvain was placed on paper, I sent it by one of my adjutants to General Ilten, who was not at the conference being incommoded and he asks to give orders to his troops to assist us to put it in execution. I went there myself with Generals Mölck, Aylva, Hompesch, and Dalwig to overcome all obstacles. Nothing being able to succeed in producing some taste for this enterprise, we agreed to march forward while awaiting these troops and to return to the execution until after the arrival of the Imperial troops." (Waldeck, 18 February).

longer, as long as he could. I communicated this news to the generals and I made dispositions to march the following day, according to what had been arranged."[692]

The weather was pressing in effect. On arriving on the 20th, at Malines, where the Prince had transported his headquarters, Zastrow had been killed the day before the siege cannonade. An emissary from van der Duyn arrived immediately afterwards, saying that in Brussels "they were occupied with capitulating." The following day a new messenger, a major in the Prince's own regiment, came to completely destroy his last hopes. He carried the definitive account of the capitulation. "I had" said the Prince, "the satisfaction of learning that all the garrison, following the example of their general, had demonstrated all the valor and good will possible and that the city, which was not a place of war, was only surrendered after having been breached."[693] The zeal and activity displayed by von Waldeck merited better than this consolation that was comforting, but superfluous.

Admirably informed by Löwendahl, Maurice de Saxe was not unaware of the arrival of the Imperial troops. Consequently, he could scarcely allow himself to be amused by the maneuvers executed by the Prince and van der Duyn. It was not long before he saw that the Allies "did not act in good faith" during the negotiations undertaken since the raising of the white flag.[694] Maurice de Saxe showed an unshakable resolution to not abandon the principal stipulation that the garrison become prisoners of war. The Allied generals spoke of not accepting this hard condition, claiming they were certain to be relieved. If one is to believe d'Espagnac, Maurice responded: "Well, gentlemen, it is only (men without hearts) that surrender when they expect to be relieved; return to your walls and defend yourselves there."[695] The report on the negotiations addressed by Maurice to d'Argenson lets us believe that the Marshal was perhaps less categorical, but that his habitual manner showed him to be very skillful and also very prudent.[696]

---

[692] Waldeck, 18 & 19 February.

[693] Waldeck, 20 & 21 February.

[694] Maurice de Saxe to d'Argenson, headquarters at Laken, 21 February (Coresp. A.F.).

[695] *D'Espagnac* (Vol. II, p. 151) *loc. cit.* – The expression placed by d'Espagnac between the parenthesis it replaces, without a doubt is another more energetic, but less proper word.

[696] Maurice de Saxe to d'Argenson, headquarters at Laken, 21 February (Coresp. A.F.).

Maurice De Saxe's 1745 Campaign in Belgium

The Maréchal de Saxe signed two capitulations, one with von Kaunitz for the Austrians and the other with van der Duyn relative to the Dutch. The French captured 18 battalions and 9 squadrons from the Dutch and 900 Austrians, forming a total of almost 15,000 men, plus 18 Austrian generals, 2 Dutch generals, and 100 officers of all grades who were in Brussels for personal affairs. Von Kaunitz, who was not a soldier, retained his liberty. In Brussels and Vilvorde the French captured 108 cannon, including almost the entire Dutch field artillery park, 30 pontoons, 52 flags and 3 standards.[697]

From 28 January to 22 February, the French infantry, alone employed in the siege, suffered 293 dead and 639 wounded. A total of 492 sick remained in hospital. There were only 228 soldiers who had deserted or disappeared.[698] The French lost 9 officers killed and 37 wounded.[699]

The French found, in the Brussels hospitals, 500 wounded Dutchmen, and in the magazines only 300,000 rations of forage and 150,000 of oats. This was an insufficient supply. The Marshal requisitioned immediately, from the States of Brabant an equal supply. They saw in this obligation to draw this from the region northeast, as far as Louvain, the area around the city being empty. Also, the Allies found themselves, by this measure, deprived of part of their own subsistence. The French found, in Brussels, no magazine of wheat or flour. The Dutch troops lived on their pay; the military administration did not produce their bread, at least in the fortresses and during winter quarters. As a result, it was necessary to bring, from Ghent, 4,000 sacks of flour.[700]

On 25 February, three divisions of Dutch prisoners left for France. The Court came finally to find the occasion to exercise the reprisals so long anticipated. It had informed the Marshal of the Royal will. "The states should not flatter themselves to quickly see the return of their 18 battalions, after the example that they gave us of their lack

---

[697] State of troops that formed the garrison of Brussels and became prisoners of war (Corresp. A.F.).

[698] State of the soldiers killed or wounded during the siege of Brussels, those in hospital, and those missing from the organization of the army to 22 February (*Ibid.*). – This situation was established in a manner to give the details for the 18 regiments, including the artillery, effectively employed in the siege.

[699] State of officers killed or wounded during the siege of Brussels (*Ibid.*). – This situation is nominative and, as the two preceding, it is attached to the original copies of the two capitulations signed by the Marshal, von Kaunitz, and van der Duyn.

[700] Séchelles to d'Argenson, Brussels, 23 February (Corresp. A.F.).

Henry Pichat

of faith in the most solemn and most respectable engagements [they had given us]," wrote d'Argenson.[701] Since the 12th, and for greater precautions, the Minister had already given Maurice and Séchelles definitive orders relative to the treatment of the prisoners "that we actually have and to those which will be subsequently captured."[702] Very detailed instructions were left to the officers charged with escorting these captives, as well as to the Intendants of the provinces that these columns would cross.[703] These divisions were sent to Laon, Amiens, and Noyon, while preparations were made to send them into the Orléanais, Touraine, Anjoy, and Berry. The Austrians left for Mons on the 27th. The weapons of the Allied garrison of Brussels, assembled in a depot in the city, would not be returned to the Dutch until the signing of a general peace.

In all the letters addressed to the court since 5 December, the Marshal had not ceased to advise against the occupation of Brussels. The Minister had given him no response until 25 December. However, after that date, when the landing on England appeared to be indefinitely postponed, d'Argenson seemed to direct a new activity to the study of the Marshal's plans. He expressed some regret taken by Maurice to not occupy Brussels. He thought that perhaps it would have been preferable to maintain a French presence in the city so as to deny the Allies any possibility of returning down the Vilvorde Canal. Nonetheless, he submitted to the wisdom of the Marshal, since the latter judged it sufficient to dismantle the defenses of Brussels.[704]

Upon the departure of the expedition the ideas of the Court appeared to be modified. D'Argenson insisted anew that the Marshal occupy Brussels. The Minister estimated that the occupation of the city would not be without utility when the moment arrived to open the 1746 campaign. In addition, the destruction of the defenses of Brussels would include the trenches constructed by the Allies along the canal as well as the Vilvorde sluices and the bridges. Wasn't these works useless and in ruins already?[705]

On 2 February 1746, the Minister returned at the charge with a force much greater now that the Court had already discounted the

[701] D'Argenson to Maurice de Saxe, Versailles, 24 February (*Ibid.*).

[702] D'Argenson to Maurice de Saxe and Séchelles, Versailles, 17 February (*Ibid.*).

[703] D'Argenson to Chauvelin, Méliand (and to the Intendants of Picardy, Soissons, Alençon, Tours, Bruges, Moulins, Orléans), Versailles, 27 February (*Ibid.*)

[704] D'Argenson to Maurice de Saxe, Versailles, 25 December 1745 (Corresp. A.F.).

[705]D'Argenson to Maurice de Saxe, Versailles, 25 January 1746 (*Ibid.*).

capture of Brussels. D'Argenson completely developed the rationale that he wished to pursue in the opening of the coming campaign. According to him, the occupation of Brussels was of such a nature as to throw the Allies into a cruel embarrassment. It obliged them to find other quarters for the rest of winter beyond the Meuse. It was, in effect, impossible for them to hold Louvain and Malines and they held no fortress or post between the Meuse, the Dyle, and the Demer. Under these conditions, the Austrian troops coming from the Rhine had to join the Dutch by Limburg or Maëstricht or throw themselves into Mons, Namur, and Charleroi. One could, therefore, regard the Allies as cut in the middle and separated one from the other, concluded d'Argenson.

The Minister responded, in advance, to the objections raised by the Marshal. He reported that the occupation of Brussels, Malines, and Louvain obliged almost all French troops to leave their winter quarters. He saw also that this movement uncovered Hainaut, the region between the Sambre and the Meuse, and risked leaving these regions open to the Allied garrisons at Mons, Namur, and Charleroi. Finally, the disorder imposed on the troops could harm the completion of their rebuilding. D'Argenson did not conceal the importance of all these considerations. Why, he added, did the Marshal not propose to the King using some of the battalions wintering in Évêchés to occupy Brussels, Louvain, and Malines or to cover the frontier?[706]

At this moment, there occurred an event capable of exercising a degree of influence on the decisions of the heretofore unshakable marshal to not occupy Brussels. On 3 February, de Wassenaer wrote from La Haye to the Marshal asking that he facilitate for him the means of arriving at Versailles. Wassenaer told him he was, "charged with the orders of Their High Powers to His Most Christian Majesty." He also appeared a little pressed to receive the passport that he solicited.[707] He attached to his letter some comments and considerations in which the Minister found the response to his two previous letters.

The Marshal found Wassenaer's haste most unique. "This did not at all resemble the phlegmatic Dutch," he said.[708] In addition, the very different attitudes of the plenipotentiaries and the Dutch general caused one to think that there was some discord among the members

---

[706] D'Argenson to Maurice de Saxe, Versailles, 2 February 1746 (*Ibid.*).

[707] Wassenaer to Maurice de Saxe, La Haye, 3 February (*Ibid.*).

[708] Maurice de Saxe to d'Argenson, headquarters at Laken, 6 February, 5:00 a.m. (*Ibid.*).

of the Council of the "Most Serene Republic."[709] In sum this was an important event. "This changed the question," wrote Maurice, "and if, as I suppose, in it, you demand from the Dutch a pure and simple neutrality, with the evacuation of all their troops from the Brabant, Hainaut, as well as the County of Namur, it is right to remain in Brussels." Moreover, Marshall proposed "to hasten the will of States"; he thought to convince them to treat definitively by advancing on their frontiers. For this object, he proposed to form four strong columns. The left column was to move along the canal to the Escaut; the second would march on Antwerp. The other two would march on Lierre and Campine. Also, the division of the French forces would facilitate them to lodge themselves and live there during the end of the bad seasons.

The habitual circumspection of the Marshal inspired in him, however, some fears. If de Wassernaer did not go to Versailles with the sincere intention of treating, it was necessary that he assure he not make the same error as the Allies. Why occupy, in their turn, Brussels, Malines, and Louvain, fortresses that were untenable as all the land between the Dyle and the Escaute was exhausted? Maurice feared that the Court "would cause him to make some false movement." At the same time as he transmitted his suspicions to d'Argenson, he also wrote to Noailles: "to deceive me on this point it would cause you to mislead yourself."[710]

These fears do not appear to have been in the least justified. The clearly hostile sentiments fed by the Court in regard to the Dutch soon received a most agreeable satisfaction. Brussels would fall and in Versailles one no longer showed any more haste to receive von Wassernaer than he showed to come. The King authorized the Marshal to give him a passport. However, "before giving the orders relative to von Wassenaer's mission" Louis XV proposed to confer with his council.[711]

On 12 February, d'Argenson succeeded in reassuring the Marshal, but the question of the occupation of Brussels remained open. According to the information sent from Holland to the Minister of Foreign Affairs, it was believed in Versailles that Wassenaer would present a project for a neutrality accompanied with such conditions that Louis XV could not treat with them. "You will not await the re-

---

[709] Maurice de Saxe to Bernier (Adjutant General of the Army), headquarters at Laken, 7 February (*Ibid.*).

[710] Cf. *Maurice de Saxe et le marquis d'Argenson*, (Vol. I, p. 62), *loc. cit.*

[711] D'Argenson to Maurice de Saxe, Paris, 8 February (Corresp. A.F.).

sults of von Wassenaer's negotiations to prepare your military movements," concluded d'Argenson.[712]

The letter from the Minister was received by at Maurice about the same time as d'Hérouville bearing another signed by Richelieu and coming from Calais. The duke informed the Marshal that he was returning to Paris. Richelieu had found, at the moment of embarkation, "insurmountable obstacles."[713] He turned over to Maurice the command of the troops destined for the expedition to England. "That changes the thesis," said the Marshal, "and I can begin to think of the means to holding Brussels, when I have made myself master of it, as previously I could not think of this because I was too weak to march to its succor if the enemy had besieged it in their turn."[714]

As a result, Maurice counted on leaving in Brussels the artillery that had participated in the siege, 12 battalions, a regiment of dragoons, and Löwendahl was to command it all. On 18 February, he made the necessary dispositions. He brought to Bruges and Ghent the French battalions spread by Richelieu along the shores of Picardy, planning, no doubt, to use these troops to garrison Brussels. He only left the Irish troops, until the receipt of new orders, at the disposition of Lord Clare, commander in place of Richelieu.[715]

These measures received the approval of the Court. It appeared, in effect, to have only a limited taste for the initiative taken by Richelieu. The King was unaware, at least if one believes d'Argenson, that the Duke had returned command of his troops to Maréchal de Saxe. Although Richelieu had obtained permission to return to Paris for reasons of his health, the Court had not counted on the projected plan for a landing. As a result, Maurice de Saxe stopped at Ghent and Bruges, until new orders were received, the movement of the battalions returning from Picardy and he was not to, under any pretext, move the Irish troops.[716]

The Marshal had been certain, for a long time, that the expedition to England would not occur. It is easy to read this thought between the lines of the correspondence between Maurice and d'Argenson from the month of December 1745. However, he had never taken the liberty of speaking openly about it. In contrast, he had not

[712] D'Argenson to Maurice de Saxe, Versailles, 12 February (Corresp. A.F.).
[713] Maurice de Saxe to d'Argenson, headquarters at Laken, 15 February (*Ibid.*).
[714]*Ibid.*
[715]Maurice de Saxe to d'Argenson, headquarters at Laken, 28 February (*Ibid.*).
[716] D'Argenson to Maurice de Saxe, Versailles, 20 February (Corresp. A.F.).

ceased to scrupulously conform to all the orders that he had received, to give his personal support or that of his troops to Richelieu. The initiative taken by the latter, however lightly, if one is to believe d'Argenson, had placed the Marshal in a disagreeable situation. Maurice did not complain. He contented himself with excusing himself for the "great fault" that he had committed in displacing the troops stationed in Picardy. He attached, however, to the expression of his regrets, the letter that he had received from Richelieu. This letter contained nothing "but politenesses", it is true, but from the departure of the Duke, Lord Clare, adjutant to Richelieu, had sent another that the Marshal anxious to focus things, attached to the first. Clare gave many more details than his chief. The reading of this last letter had convinced the Marshal that it was d'Argenson's to pass to Scotland, in small groups, troops that one could not send in a mass, to assist the Jacobites. In these conditions, he had thought to arrange the French troops in Picardy. The operation on Brussels and the descent on Scotland could not be pursued simultaneously. "These are the reasons that I argue to you for my justification," concluded the Marshal.[717]

It soon became certain that the very skillful actions of Maurice had closed the incident and by the same blow answer the question, not still definitively resolved, of the occupation of Brussels. On 27 February, d'Argenson sent to the Marshal a *satisfecit* concluded in very suggestive terms. It gave him full powers to complete the necessary dispositions.

From the 23rd, Maurice had sent Löwendahl to Brussels. In concert with the engineers, they had inspected the defenses of the fortress with an eye to determining exactly the force of a garrison that would have to be installed. It was resolved to construct some works, the fortifications being recognized as in sufficient. It was also decided to give Löwendahl the forces necessary to hold for at least a dozen days in case the Allies attempted to besiege it. That delay would permit the French to relieve Brussels.[718] If the Allies attempted a rapid attack against the place it would be difficult to pull together a siege equipage with the resources available at Namur and Mons. The operation demanded time and, in light of the season, the movement of the necessary artillery before Brussels would be very slow. It could, thus, be captured, without great risk, before it could be brought to its

---

[717] D'Argenson to Maurice de Saxe, Versailles, 27 February (*Ibid.*).
[718] Maurice de Saxe to d'Argenson, headquarters at Laken, 23 February and Brussels, 27 February (*Ibid.*).

destination. This was enough to make Löwendahl safe from surprise. It was necessary to assure the provisions for the garrison. The 4,000 sacks of wheat that Séchelles had brought, as one knows, since the capitulation, were reserved to feed Löwendahl's troops for two months. The Marshal resolved, as a result, to activate the providing of 300,000 rations of forages that it had requisitioned. "As I know their slowness and the excuses that they present, which are ordinarily the holdup that the hussars imposed on the delivery of this forage, I have begun by the execution [of it]," said Maurice.[719] He sent 1,600 horses out of Louvain and sent them to Tirlemont. The army furnished 1,000 masters who left for the region of Namur. This cavalry requisitioned all the wagons of the region, loaded them with the forage that they found, and returned to Brussels, on the 29th. They brought with them the mayors and the bailiffs as hostages until the provision of the contribution imposed on the States was complete.[720]

From the first days of March, the French troops marched out to regain the cantonments they had left before the definitive opening of the 1746 campaign. The movements began on 7 March with the reentry in their corps of the different detachments remaining at the posts at a distance from Brussels. On the 3rd, du Chayla, who had replaced Löwendahl in the command of Louvain returned to Brussels. This concentration of the army around the fortress had as its goal reconstituting the six divisions. The troops were to, in principle, regain their former cantonments in which the regiments had left their depots.

On the 4th, the first separation occurred. Brézé moved on Hal with the troops that he was going to take to Tournai. They were joined there with those commanded by Relingue, but which Phélippes had brought from Hainaut. The following day there occurred a general movement. Clermont-Gallerande left for Ninove. Brézé continued by Soignes along the road he had followed the day before. D'Armentières marched on Enghien, de Vaux marched directly on Dendermonde and the Marshal, with his Swedish artillery, moved on Alost. During these marches, Grassin, with 2,000 men and some miners destroyed the gates of Nivelle and dismantled the walls of this small fortress.[721] Having a difficult access, it had to that point constituted to that time an excellent lair for the free troops that von Waldeck had installed there in October 1745.

---

[719] Maurice de Saxe to d'Argenson, Brussels, 27 February (Corresp. A.F.).
[720]*Ibid.*
[721]Order to return the troops to their quarters (de Vault, Piece 79).

Henry Pichat

Along the Brussels Canal, d'Hérouville leveled the works and demolished the bridges established by the Allies since July 1745. After these demolitions, only men on foot crossing the narrow planks of the sluices or by small boats could penetrate into the Brussels region. The 600 men established in Alost assured the communications between Brussels and Ghent.

Brézé had received full authority to explode the walls of Hal, Soignes, and Braine-le-Comte in order to remove all support for parties from Mons, Namur, and Charleroi. It appeared to him sufficient to break numerous breaches in the walls of the convent neighboring Braine-le-Comte which constituted a much more serious post than the city itself.

These precautions sufficed to put all the conquered territory safe from the enterprises from the Allied pillagers for the two months that remained until the opening of the coming campaign. On 8 March, all the French troops had returned to their cantonments. On the 9th, the Marshal left for the Court. He arrived there on the 12th and was greeted in Paris enthusiastically. Before his departure he had given overall command to Clermont-Gallerande, the most senior of the lieutenant generals present in Flanders. He left him detailed instructions to put into execution at the least indication of an Allied march on Brussels.[722]

In its cantonments, the French army contained 126 battalions and 164 squadrons divided into 12 commands, as follows:

1 – Brussels, Fort Monterey and Vilvorde: 15 battalions, 5 squadrons under von Löwendahl.

2 – Ghent, Thielt, Alost, Dendermonde: 25 battalions and 35 squadrons under du Chayla.

3 – Bruges, Damme, Ostend, Nieuport: 12 battalions and 4 squadrons under the orders of de Cantades.

4 – Audenarde and from that city to Ghent: 4 battalions and 13 squadrons, commanded by Clermont-Gallerande.

5 – Tournai: 9 battalions and 8 squadrons under de Brézé.

6 – Ath: 5 battalions, 7 squadrons under d'Armentières.

7 – Dunkirk, Calais, Gravelines, Boulogne, Bergues, Montreuil: 10 battalions and 9 squadrons under d'Aunay.

8 – Ypres, Dadizeele, Bacelaere, Wervike, Commines, Warneton, Lille, Béthune, Saint-Venant, Aire, Saint-

[722] De Vault, p. 170 and following.

Omer, Arras: 9 battalions and 24 squadrons under de Ceberet.

9 – Courtrai, Menin, Haerlebeke, Deinse: 10 squadrons under de Montmorency-Logny.

10 – Valenciennes, Condé, Douai, Orchies, Bouchain, Cambrai, Lequesnoy: 16 battalions and 16 squadrons under Le Danois.

11 – Maubeuge, Landrecies, Avesnes, Beaumont, Philippeville, Givet, Mariembourg: 15 battalions and 17 squadrons under de Phélippes.

12 – 6 battalions and 4 squadrons, cantoned in Boulogne, Gravelines, Dunkirk, and Ostende and commanded by Lord Clare, under the orders of Richelieu, forming a command in part. Finally to have the complete strength of the French forces, it is necessary to add to the preceding 12 squadrons of cavalry, which after having escorted the three divisions of Dutch prisoners, was to subsequently rejoin the main army at Saint-Omer, Aire, and Béthune.[723]

---

[723] De Vault, Piece 78. – See the details in the annexed documents.

Henry Pichat

ANNEXED DOCUMENTS

General State of the Army of Flanders
1 July 1745

Army of the King

	Battalions		Battalions
Gardes françaises	6	Nivernois	1
Gardes suisses	3	Rohan	1
Picardie	4	Dauphiné	1
Piémont	4	Fleury	1
Normandie	4	Royal-Corse	1
Crillon	3	Beauvoisis	1
Laval	1	Courten (Swiss)	3
Auvergne	3	Bettens (Swiss)	2
Royal-Lorraine	1	Wittemer (Swiss)	2
Bouzols	3	Seedorf (Swiss)	3
Biron	1	La-Cour-au-Chantre (Swiss)	3
Le roi	4	Monin (Swiss)	3
Royal	3	Diesbach (Swiss)	3
Hainaut	1	Bulkley (Irish)	1
Dauphine	3	Clare (Irish)	1
Traisnel	1	Dillon (Irish)	1
Gondrin	2	Rooth (Irish)	1
Angoumois	1	Berwick (Irish)	1
Soissonnois	1	Royal-Escossois (Irish)	1
Touraine	3	Lally (Irish)	1
Saintonge	1	Richecourt (artillery)	1
Eu	2	Fontenay (artillery)	1
Löwendahl	2	Pumbecq (artillery)	1/2
Royal-Vaisseau	3	La Tour (Royal Grenadiers)	1
Languedoc	1	D'Espagnac (Royal Grenadiers)	1
Orléans	2	Bouteville (Royal Grenadiers)	1
Chartres	2	Valfons (Royal Grenadiers)	1
La Couronne	3	Arquebusiers de Grassin	2

Total 108½

Cavalry

	Squadrons		Squadrons
Maison du Roi	13	Prince Camille	4
Gendarmerie	8	Orléans	4
Colonel Général	4	Bramcas	4
Brionne	4	Clermont-Prince	4
Royal	4	Harcourt	4
Rohan	4	Penthièvre	4
Le Roi	4	Noailles	4
Clermont-Tonnerre	4	Beausobre (hussars)	4
Royal-étranger	4	Linden (hussars)	4
Chabrillant	4	Saxe-volontaires (uhlans)	6
Cuirassiers	4	*Dragoons*	
Egmont	4	Mestre de camp général	5
Cravattes	4	Royal	5
Fiennes	4	Asfeld	5
Royal-Roussillon	4	Egmont	5
Saint-Jal	4	Beauffremont	5
Carabiniers	10	Septimanie	5
Berry	4	Arquebusiers de Grassin	4

Total   167

Troops under the orders of Clermont-Gallerande

La Fère	1 battalion
Bettens	1
Wittemer	1

	Total	3 Battalions
Mestre de camp général		4 squadrons
Beaucaire		4
Bellefons		4
Anjoy		4
Andlau		4
Bourbon		4
Maugiron		4
Berchiny (hussars)		6

Total   34 squadrons

Troops detached in Garrison

Löwendal, at Valenciennes		1 battalion
Royal-Wallon, at Saint-Omer & Saint-Venant		1
Royal-Wallon, at Aire and Béthune		1
Boufflers-Wallon, at Dunkirk		2
Militia battalions, 7 at Tournai &		
3 guarding the line of the Chin		10
	Total	15 Battalions

Talleyrand at Lancrecies		4 squadrons
Grammont at Givet, Marienbourg, & Philippeville		4
Fritz-James at Tournay		4
	Total	12 squadrons

French Forces at the Siege of Ostend

Commanding officer: von Löwendal

Maréchaux de camp: Count de la Marck, de Contades, d'Armentières, de Seedorf, de la Suze, d'Hèrouville.

Aides-majors gènèraux: de la Tour, d'Hallot, Michault (for the trench)

Engineers: Gourdon, Thierry, des Noyers, Biscours

East at Bruges	Crillon	3 battalions
	Laval	1
Arriving on the 4th, at Ghent,	Eu	2
the 5th at Aeltere, &	Löwendal	2
the 6th at Bruges	Bettens	2
	Wittmer	2
	Seedorf	3
	La Clour-au-Chantre	3
Arriving the 4th at Aeltere	Royal Grenadiers	4
& the 6th at Bruges	Saint-Brienne Militia	1
Date of arrival unknown	Royal-Artillerie	1
	Total	24 battalions

The 7th at Bruges	Beauffremont Dragoons	5 squadrons

French Forces at The Siege of Ath

Commanding officer: de Clermont-Gallerande
Maréchaux de camp: de la Suze, de Saint-Pern, de Bouzols, de la Vassé, de Fines, de Chépy
Brigadiers: de Vence, de Bombelles, de Laval
Staff:
    Aides-majors généraux: d'Espagnac, de la Tour, de Pradines
    Aides-maréchaux de logis de la cavalerie: de Mouchy, Chevaler de Mesières
    Commissaire Ordonnateur: de Crance
    Commissaire ordinaire: Foulon

	Battalions		Battalions
Picardie	4	Seedorf	3
Auvergne	3	Dauphiné	1
Gondrin	2	Royal-Corse	1
Rohan	1	Löwendal	2[724]
Laval	1	Touraine	3
Langauedoc	1	La Fère	1
Bettens	3	Nivernais	1
Saintonge	1	Hainaut	1
Soissonais	1		

	Artillery		
Pumbecque	½	Richecourt	1[725]

Cavalry
Under the orders of de Clermont Gallerande

	Squadrons		Squadrons
Anjoy	4	Brancas	4
Orléans	4	Andla	4
Bourbon	4	Maugiron	4
Beaucaire	4	Royal dragoons	5

Under orders of d'Estrées

Le Roi	4	Egmont	5
Clermont-Tonnere	4	Asfeld	5
Mestre de camp général	5	Beauffremont	
		Dragoons	5

---

[724] The third battalion was at Valenciennes
[725] Coming from Tournai.

Maurice De Saxe's 1745 Campaign in Belgium
Emplacements of French Troops in Siege of Brussels

Lieutenant General von Löwendal
Maréchal de camp d'Avarey

Emplacement	Regiments	Battalions	Squadrons
Brussels	Piedmont	4	-
	Dauphin	4	-
	Les Vaisseaux	3	-
Fort de Monterey	Diesbach	3	-
& Vilvorde	Det/Fontenay (artillery)		-
	Chartres	2	-
	Mestre de camp général dragons		5
	200 Grassins	3	-
	Total	15	5

Lieutenant General du Chayla
Maréchal de camp d'Hérouville

Emplacement	Regiments	Battalions	Squadrons
Ghent & Ghent Castle	Normandie	4	-
	Royal	3	-
	Limonsin	3	-
	La Couronne	3	-
	Beauvoisis	1	-
	Rochefort	1	-
	La Cour-au-Chantre	3	-
	Vire (militia)	1	-
	Saint-Denis (militia)	1	-
	Montargis	1	-
	Fontenay (artillery)	1	-
	Royal		4
	Du Roi		4
	Cuirassiers		4
	Royal-Roussillon		4
	Saxe-Volontaire (uhlans)		6
Vicinity of Ghent	Cravattes	-	4
Thielt	Rohan	-	4
Alost	950 men detached		
	from Ghent &400 horse		
Dendermonde	Traisnel	1	-
	Angoumois	1	-
	Vernon (militia)	1	-
	Asfeld Dragoons	-	5
	Total	25	35

Henry Pichat

Lieutenant General de Contades
Maréchal de camp de la Motte-d'Hugues
Lieutenant General de Phélippes
Maréchaux de camp: Seedorf, Gravel, la Motte Guèrin & Relingue

Emplacement	Regiments	Battalions	Squadrons
Bruges	Crillon	3	-
	Eu	2	-
	Nantes (militia)	1	-
	Corbeil	1	-
	Berry	-	4
Damme	Detachment from Bruges militia	-	-
	1st Battalion, Orléans	1	-
Ostend	Rouen (militia)	1	-
	Joigny (militia)	1	-
Nieuport	2nd Battalion, Orléans	1	-
	Saint-Briene (militia)	1	-
	Total	12	4

Lieutenant General de Clermont-Gallerande
Maréchal de camp de Beauffremont

Emplacement	Regiments	Battalions	Squadrons
Audenarde	Monin	3	-
	3rd Battalion, Paris (militia)	1	-
	Beauffremont Dragoons	-	5
Vicinity of Audenarde	Beausobre (hussars)	-	4
Between Audenarde & Ghent	Brienne	-	4
	Total	4	13

Lieutenant General le Danois

Emplacement	Regiments	Battalions	Squadrons
Valenciennes & Citadel	Le Roi	4	-
	Amiens (militia)	2	-
	1st & 3rd Battalions, Ligneville (militia)	1	-
	Bourbon	-	4
	Rohan	1	-
Condé	Chartres (militia)	1	-
Douai	1st Battalion, Paris (militia)	1	-
	Chaumont (militia)	1	-
	Orléans	-	2

Maurice De Saxe's 1745 Campaign in Belgium

Emplacement	Regiments	Battalions	Squadrons
Orchies	Orléans	-	2
Cambrai	Poitiers (3rd Militia Company)	1	-
	Beaucaiare	-	3
Le Quesnoy	Bouffliers-Wallon	2	-
	2nd Battalion, Paris (militia)	1	-
	Troyes (militia)	1	-
	Anjou	-	4
Bouchain	Poitiers (3rd Militia Company)	-	-
	Beaucaire	-	1
	Total	16	16

Lieutenant General de Phélippes
Maréchaux de camp: Seedorf, Gravel, la Motte Guèrin & Relingue

Emplacement	Regiments	Battalions	Squadrons
Maubeuge	Fleury	1	-
	Royal-Wallon	2	-
	Orléans (militia)	1	-
	Argentan (militia)	1	-
	Mazarin (militia) (5 companies)	1	-
	Egmont Dragoons	-	5
Landreices	Hainaut	1	-
	Brancas	-	1
Avesnes	Brancas	-	1
	Metz (militia)	1	-
Beaumont	Seedorf	1	-
	Mazarin (militia) (4 companies)	-	-
	Mestre de camp	-	1
Philippeville	Seedorf	1	-
	Rennes (militia)	1	-
	Mestre de camp	1	3
Givet &	Seedorf	1	-
Charlemont	Abbeville (militia)	1	1
	Verdun (militia)	1	-
	Bourg-en-Bresse (militia)	1	-
	Talleyrand	-	3
Mariembourg	Talleyrand	-	1
	Total	15	17

General Recapitulation

	Battalions	Squadrons
Under the orders of	von Löwendal	
du Chayla		

Lieutenant General de Brézé
Maréchal de camp de la Suze

Emplacement	Regiments	Battalions	Squadrons
Tournai	Royal la Marine	1	-
	Bettens	3	-
	Wittemer	3	-
	Saint-Maixent (militia)	1	-
	Provins	1	-
	Colonel Général		4
	Fiennes		4
	Total	9	8

Maréchal de camp d'Armentières

Emplacement	Regiments	Battalions	Squadrons
Ath	Languedoc	1	-
	Nantes (militia)	1	-
	Tours (militia)	1	-
	Grassins	2	2
	Royal Dragoons	-	5
	Total	5	7

Lieutenant General Count d'Aunay

Emplacement	Regiments	Battalions	Squadrons
Dunkirk	Courten	1	-
	Vannes (militia)	1	-
	Le Mans (militia)	1	1
	Total	3	1

Regiments that were used as escorts for the Dutch prisoners taken when Brussels fell and the sites to which they subsequently went:

Emplacement	Regiments	Battalions	Squadrons
To Amiens	Chabrillant	-	4
To Noyon	Clermont-Tonnere	-	4
To Laon	Maugiron	-	4
	Total	-	12

Maurice De Saxe's 1745 Campaign in Belgium

General Recapulation

		Battalions	Squadrons
Under orders of	von Löwendal	15	5
	du Chayla	25	35
	du Contades	12	4
	de Clermont-Gallerande	4	13
	le Danois	16	16
	de Phélippes	15	17
	de Brézé	9	8
	d'Armentières	5	7
	d'Aunay	10	9
	de Cebere	9	24
	de Montmorency-Logny	-	10
	de Clare	4	12
Prisoner escort		-	12
	Total	126	164

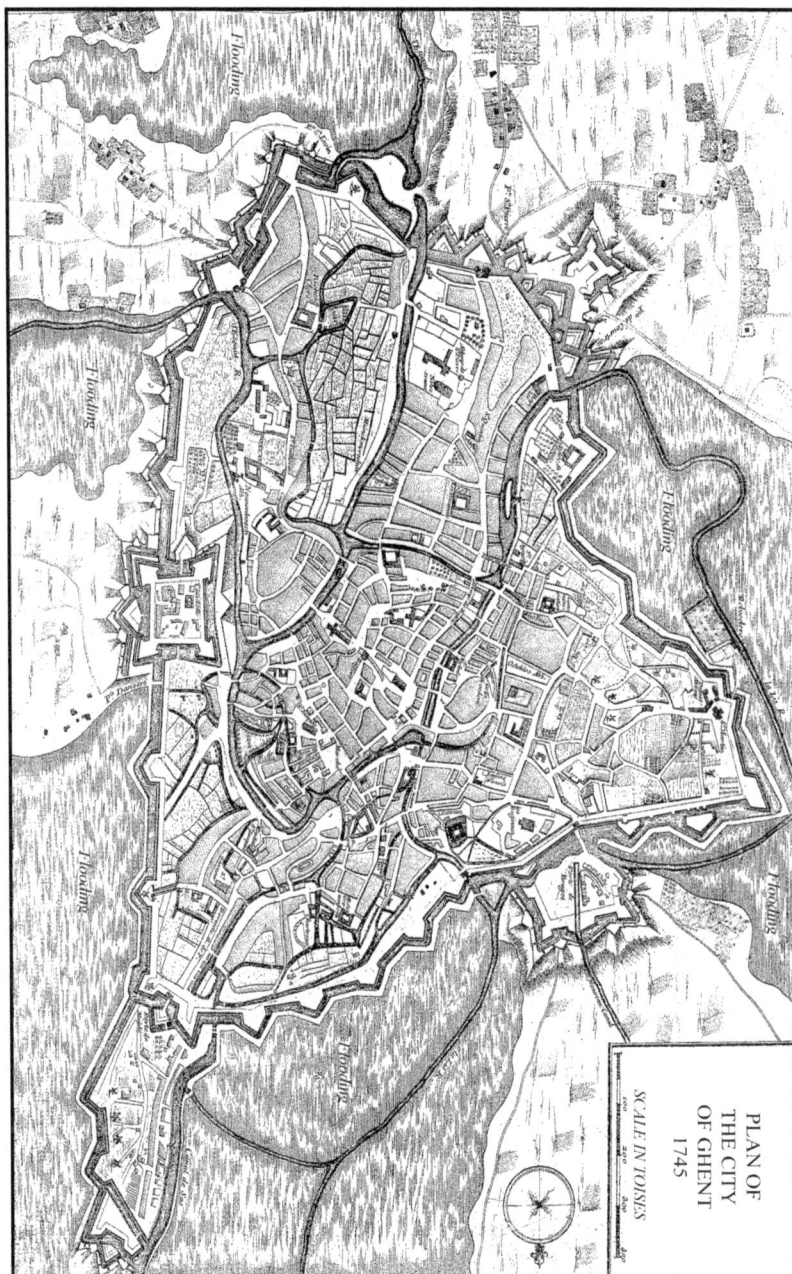

PLAN OF
THE CITY
OF GHENT
1745

SCALE IN TOISES

Plan of the Attack on Ath

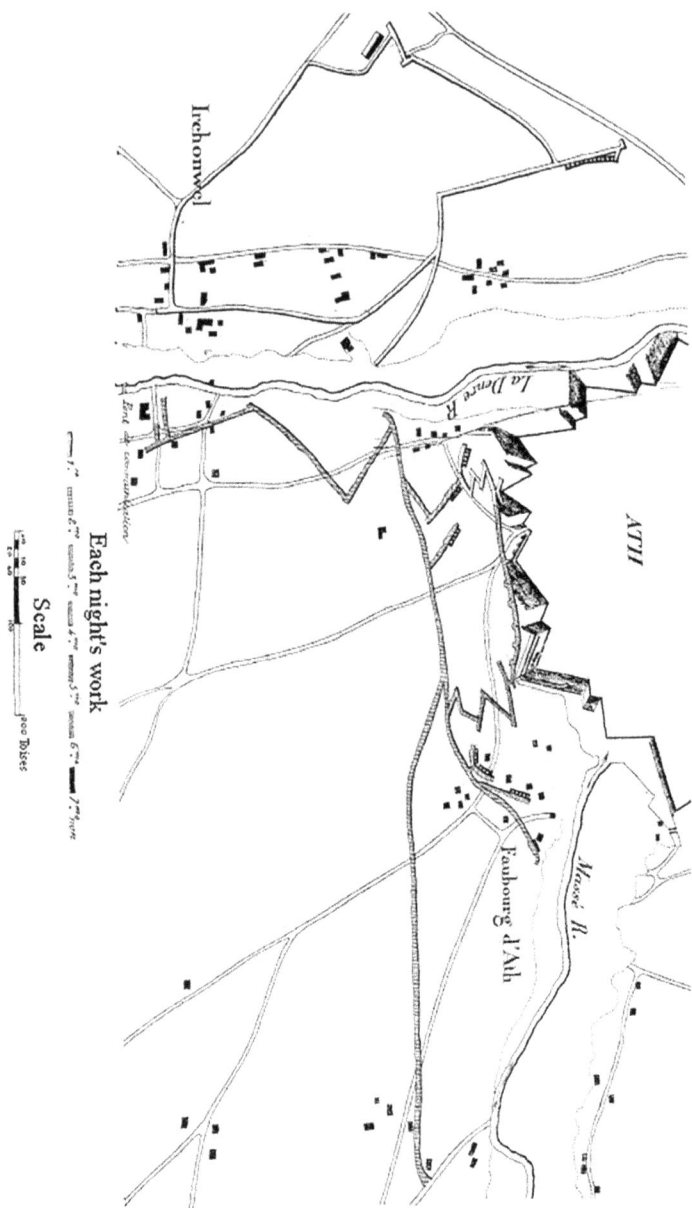

Irchonwel

La Dence R

ATH

End de circonvallation

Faubourg d'Ath

Maxvi R.

Scale

Each night's work

600 Toises

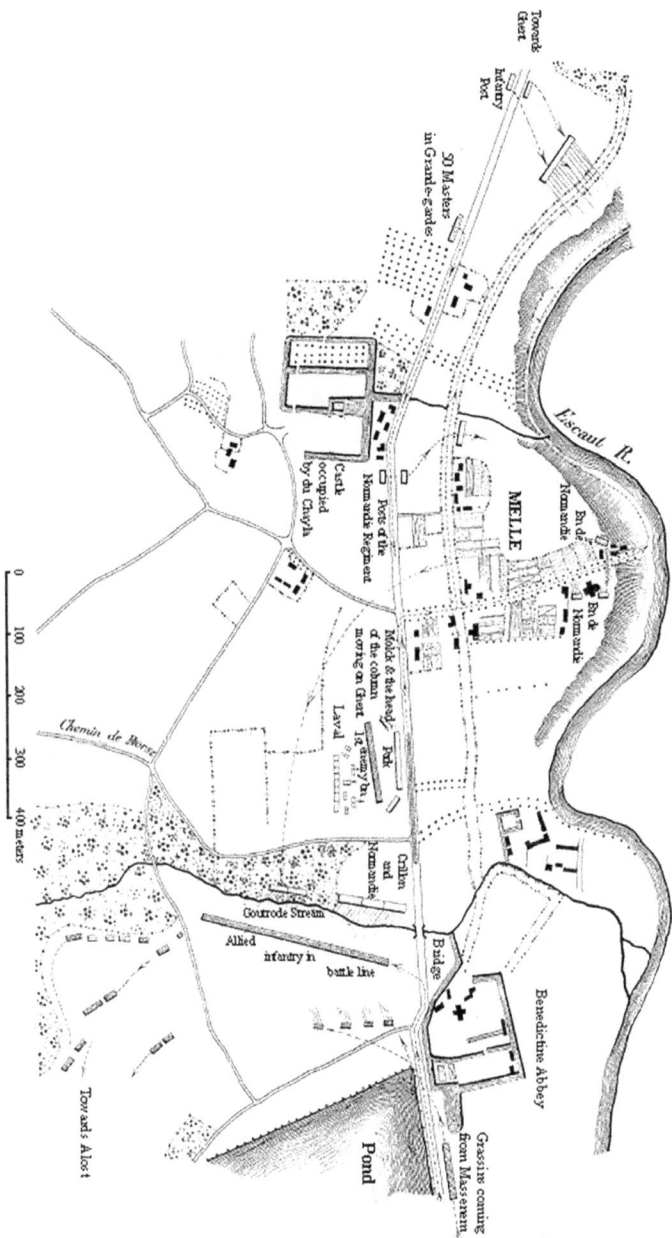

Engagement at Melle

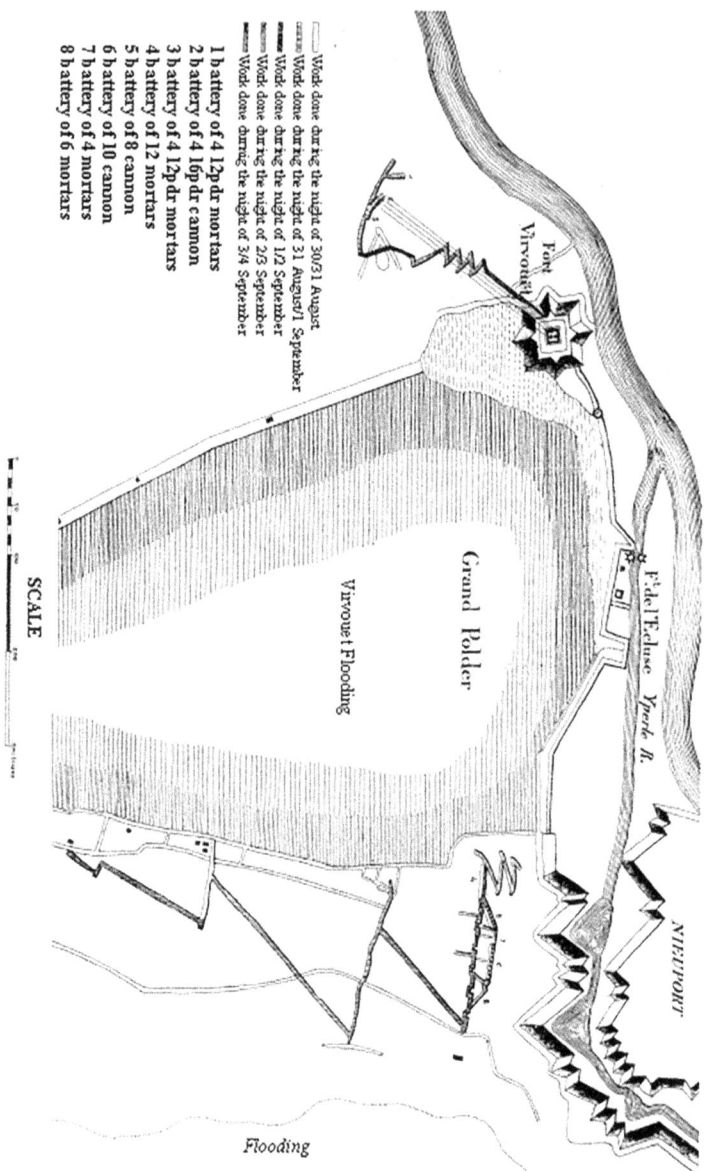

ATTACK ON NIEUPORT

Work done during the night of 30/31 August
Work done during the night of 31 August/1 September
Work done during the night of 1/2 September
Work done during the night of 2/3 September
Work done during the night of 3/4 September

1 battery of 4 12pdr mortars
2 battery of 4 16pdr cannon
3 battery of 4 12pdr mortars
4 battery of 12 mortars
5 battery of 8 cannon
6 battery of 10 cannon
7 battery of 4 mortars
8 battery of 6 mortars

SCALE

Fort
Virvotiet

Virvouet Flooding

Grand Polder

F.t del Ecluse

Ypreds R.

NIEUPORT

Flooding

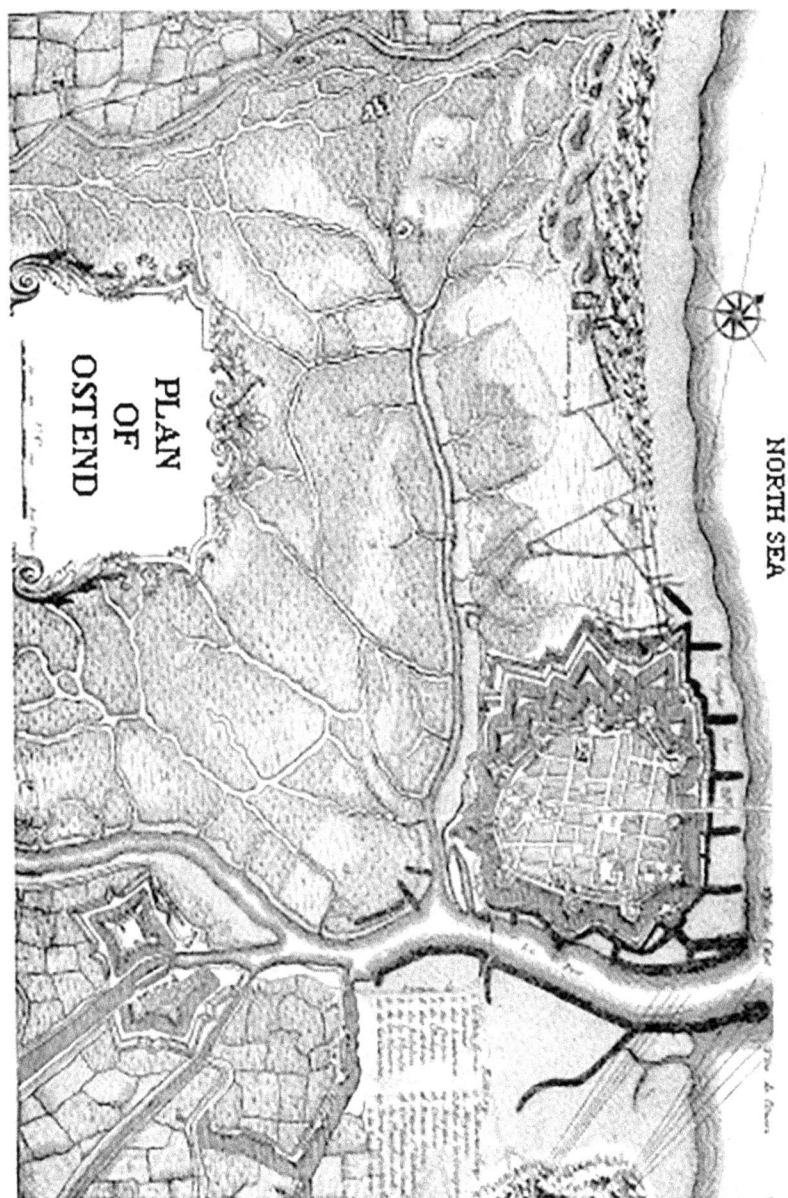

PLAN OF OSTEND

NORTH SEA

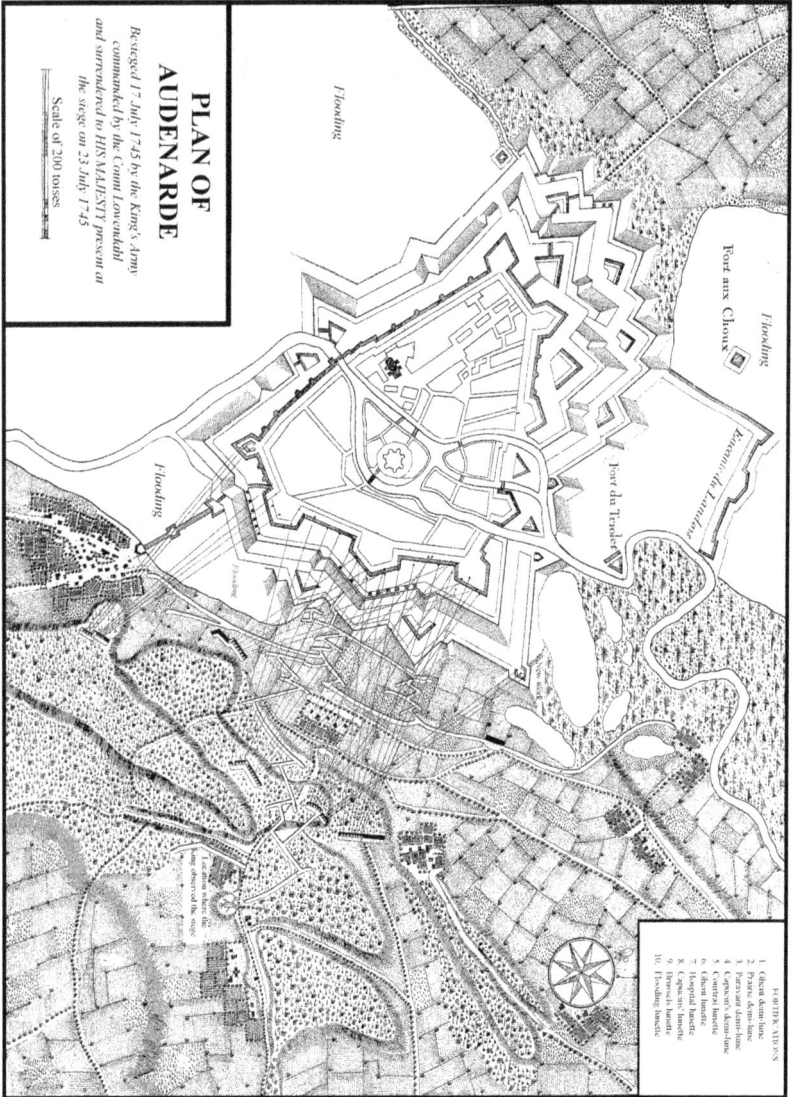

PLAN OF AUDENARDE

Besieged 1ˢᵗ July 1745 by the King's Army commanded by the Count Lowendahl and surrendered to HIS MAJESTY present at the siege on 23 July 1745

Scale of 200 toises

ATTACK ON DENDERMONDE

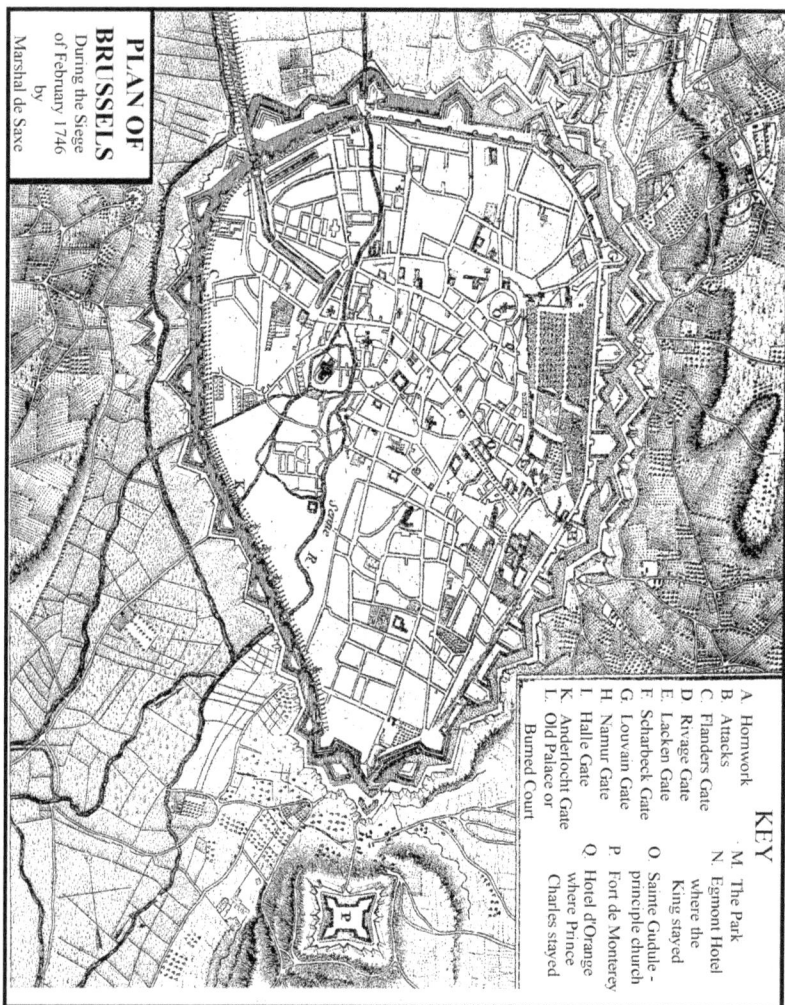

PLAN OF
BRUSSELS
During the Siege
of February 1746
by
Marshal de Saxe

KEY

A Hornwork
B Attacks
C Flanders Gate
D Rivage Gate
E Lacken Gate
F Scharbeck Gate
G Louvain Gate
H Namur Gate
I Halle Gate
K Anderlecht Gate
L Old Palace or
    Burned Court

M The Park
N. Egmont Hotel
    where the
    King stayed
O Sainte Gudule -
    principle church
P Fort de Monterey
Q Hotel d'Orange
    where Prince
    Charles stayed

BIBLIOGRAPHY

I Manuscript Sources

1. Historical Archives of the Ministry of War, Section Ancienne: Series I (Correspondence), Volumes 3084, 3085, 3086, 3087, 3088, 3089, 3090, 3091, 3092, 3098, 3099, 3100, and 17 and 18 (these last two are from the Swedish Collection) and Series V (Administration), Volumes 3123, 3125, and 3182.

The documents included in these volumes come from various sources, but they are all relative to the 1745 campaign in Flanders and Germany.

Some are the minutes of letters, circulars, orders, or specific instructions sent by the Secretary of State to the War Department for the service of all the officers or diverse agents under its authority and employed in the armies on the frontier.

The others form the complement to the preceding. These are the documents coming from the personages indicated above and addressed to the Minister: campaign plans or reports transmitted by the senior command, minutes of orders given by this latter during the course of the campaign, reports on all the operations, reports of the agents employed in the Intelligence Service, returns of the forces under arms and in hospital, minutes of capitulations, inventories made after the occupation of captured fortresses, reports relative to the administrative organization, the subsistence of the Army and its detachments, the raising of pioneers, prisoners of war, temporary hospitals, contributions extracted, on food and material supply convoys, and on the installation and the life of the troops in winter quarters.

Finally, these volumes contain letters addressed to the minster by his personal friends in the Army or the garrisons as well as reports and plans sent by correspondents desiring to manifest their zeal or to solicit bonuses. These documents give the details suggesting that the official relations do not always manifest and, thus, they frequently constitute precious commentaries on the latter. In addition, they singularly facilitate the study of the military ideas and the customs of the time.

These pieces enclosed in the cited documents are, for the most part, original documents. We have organized them into two general classifications: *Correspondence from the Army of Flanders* and *Cor-*

Maurice De Saxe's 1745 Campaign in Belgium
respondence from the Army of the Rhine.

It is, otherwise, easy to find them in the various volumes of this collection. Each volume mentions, in effect, Flanders or Germany with the instructions for the period to which the documents it contains belong. In addition, an analytical table of the materials is almost always placed at the head of each volume.

2. Historical Archives of the Ministry of War, Section Ancienne, Series IV: Work of the King.

Each carton of this series bears the date of the documents it contains. These latter are reports that are more or less summaries submitted to the King by the Secretary of State to the Department of War. They report on the affairs which are their source. All these documents bear the decision taken by the Monarch. They do not concern only particular or personal affairs, such as requests and titles to the subject of pensions, bonuses, nominations, propositions, promotions, and letters of special services; they treat, thus, with measures of a general order: projects of regulation, of organization, or the modification for the structure of troops, particular corps, or services. The *Work of the King* is, in sum, the indispensable complement of the correspondence of the Minister.

3. Historical Archives of the Ministry of War, Section Ancienne, Series II (Historical Reports), Vol. 3093: Reports or Extracts from the Correspondence of the Courts and the Generals, 1745 Campaign in Flanders.

This volume is part of the *Vault Collection.*[726] It consists of two parts. The first, is a relation pure and simple, but detailed and complete, on the campaign, which is augmented by copies of the documents coming from the *Correspondence of the Army of Flanders,* and which M. de Vault had inserted without interrupting the account that they complete or clarify.

The second part contains only supporting documents: states of the Army or various detachments, reports, instructions, etc.; which did not find a place in this account.

[726] On the subject of this important series of reports, see D. Huguenin: *Le lieutenant général de Vault et ses Mémoires: Journal des Sciences militaries*, July 1872.

Henry Pichat

The work of M. de Vault is, therefore, above all an analytical work that is easy to verify or to complete with the original documents found in the *Correspondence of the Army of Flanders*. It permits one to make, in all confidence, use of the copies of documents where the originals have disappeared. Lieutenant General de Vault, Director of the Depot of War from 1761 to 1790, covers the armies from 1733 to 1762.

4. Historical Archives of the Ministry of War, Section Ancienne, Series II (Historical Reports): Reports written in the Month of December 1745 on the King's Campaign.

The reader of this very curious manuscript will conclude that its author, who remains anonymous, is, no doubt, an officer of a fairly high rank who participated in the 1745 campaign with the Army of Flanders. He was, in addition, a friend of many of the general officers who served under Maurice de Saxe. Finally, he was the fervent supporter of Count d'Estrées, to whom he did not hesitate to grant the honors of the day of Fontenoy.

This anonymous author did not participate in all of the actions of which he speaks. Otherwise, he conceals nothing. He is, by contrast, easy to follow in everything in which he participated, and these latter are, in truth, numerous and important. By virtue of the precise details that he frequently gives, his anonymity becomes quite transparent. We believe the author to be Louis Paul, Chevalier and later the Marquis de Brancas, who was at this time the regimental colonel of the Brancas Cavalry Regiment.

If, in effect, one consults the service records of this personage (Pinard, *Chronologie militaire,* Vol. V, p. 670), one notes that Louis Paul Chevalier de Brancas obtained, in 1739, the Brancas Cavalry Regiment, which he held until 1748. He led it in Flanders in 1744, and then in 1745 at Fontenoy, as well as at the sieges of Tournai, Audenarde, Dendermonde, and Ath, even though he became a brigadier on 1 May 1745. On the other part, the various orders of battle of the army of Maurice de Saxe mentions also the presence of the Brancas Regiment in the actions listed above. Finally, in a great many circumstances, the author was part of the detachments formed by certain brigades which he enumerates and among which, by virtue of the documents found in the *Correspondance de l'armée des Flandres*, one also finds the Brancas Regiment. The author becomes quite precise when he recounts the investment of Ath, where he occupied with "his squadrons" the Renard Woods. If one refers to the division of the troops during the circumvallation of Ath, sent to the Minister by the chief of staff of this siege corps, one observes the presence of the Brancas Regiment in the emplacements indicated by the commentator.

This manuscript is important. It gives numerous details which the author witnessed or of which he had certain information. In addition, Maurice de Saxe, Löwendal, and the others still there are alternately the object of sharp criticisms or undisguised admiration. The operations are also severely judged. The author shows, in general, little good will; however, he never fails to enumerate, at length, the motives that determined his sentiments.

For these reasons, we do not hesitate to give frequent extracts from this written account in a frequently original, lively, and sometimes acerbic, but always correct manner.

5. Historical Archives of the Ministry of War, Section Ancienne, Series I, Vol. XLIII of the Swedish Collection: Journal of the Campaign of 1745 in Flanders.

This is the complete relation of operations from 1 May to the dislocation of the French army in October 1745. One also finds tables of the composition there of the army, of certain detachments, and of the winter quarters. Its author is Michael Dreux, Marquis de Brézé (1700-1754). He has inserted in his manuscript the copy of all the march orders prepared by the staff of Maurice de Saxe. These pieces replace the originals that have disappeared from the *Correspondance de l'armée des Flandres*. One can give them as much confidence as the originals conserved in the *Correspondance* which are perfectly reproduced by the copies of M. de Bréze.

As the author of another journal cited later (see 6.) M. de Bréze has rendered his own manuscript in an epistolary form. He addresses his letters to an imaginary correspondent with clarity, sobriety, and precision. He abstains from any personal appreciation on the men and on the events, which gives his work a character of great impartiality.

Moreover, the situation of M. de Brézé at that time makes of him a well-informed chronicler. He became a lieutenant general in 1744, filling the functions of maréchal général des logis, that is to say, "chief-of-staff," to Maurice de Saxe. He then commanded at Lille during the winter of 1744 to 1745. He was employed as a lieutenant general in the Army of Flanders in 1745, serving at the siege of Tournai, and the King gave him command of that fortress on 24 May. He remained there until February 1746 when he left to conduct the siege operations of Brussels under the direction of the Maréchal de Saxe. M. de Brézé occupied, therefore, for a period of about eight months, one of the most important posts of the army in Tournai, which became the head of the line of communications for supplies and the siege depots of a great part of the troops in the field in the Lowlands.

6. Library of the Ministry of War, Manuscript A² C157: Campaigns of Flanders.

Charles de Lorraine, Prince de Pons (1696-1755) is the author of this manuscript. In 1745, he was in Flanders with the army of Maurice de Saxe. He filled the functions of his grade of lieutenant general and occasionally he was given specific commands. His journal is that of a man of the military trade. The Prince added to his manuscript 30 sketches. They are, for the most part, plans of the successive camps of the army. These plans, executed with much clarity and in a very large scale, do not fail to provide precious indications of the topography of certain parts of the Lowlands in the 18th century. They are accompanied by orders of battle and march, as well as itineraries on the columns placed under the command of Charles de Lorraine.

The journal, properly said, or account of military events, begins on 1 May 1744 and ends on 14 September 1747. To write it, the author chooses a form very much in favor in the 18th century. In this manuscript, the preface states, "was made in separate letters to avoid the constraint of a historical journal which requires the particular style of a history in which it is necessary to remove many details to preserve only the facts drawn together and led by the sequence that they have with the causes that produced them. Its epistolary style is free of this embarrassment. It must be simple and familiar. Each letter is a finished work and is a separate morsel. Premises are repeated; conjectures are ventured; one is not wrong if one is misled. Here are the reasons which have engaged me to use this method of writing."

There are numerous omissions in this work. For the year 1745 in particular, the account stops on 16 September because the Prince left the army on that date. Charles de Lorrain, in effect, concentrates on the events of which he was a witness. For the others, he does not speak of them except when he has received certain information. It is thus, by example, that he remains mute on the siege of Berg-op-Zoom, even though it is one of the most important operations, because he could not, so he says, procure "exact details." Finally, in the course of this voluminous work, he forces himself to remain strictly faithful to the promise that he had to report the facts "without praising or blaming anyone and in the most exact truth."

These scruples make Charles de Lorraine Prince de Pons a chronicler that one must not neglect.

7. Arsenal Library, Manuscript 4073: Reports of Maréchal Count de Löwendahl.

These reports begin with a long genealogical study of the Löwendahl family and ends with the Peace of Aix-la-Chapelle [Aachen]. They constitute above all a laudatory chronology but are insufficiently detailed. In addition, they offer little precision, and it is agreed that they must be used with caution. For a study of the 1745 campaign in Flanders, they make no contributions, even though Löwendahl played a very important role in it. They are far from enclosing the numerous details and precision that one finds among the voluminous correspondence of the Marshal conserved in the Historical Archives of the Ministry of War.

These manuscripts were provided by the library of the Marquis de Paulmy. It is the object of a note inserted in the catalog established by the latter. According to this note, the author of these reports was Legrand, the secretary of the Marshal, later commissioner of war. Legrand drew from *Vie du mar*échal *de Löwendahl,* by Ranft, printed in Germany, at Leipzig, in 1750, the part of these reports that he issued upon Löwendahl's entrance into French service in 1743. The Marquis de Paulmy adds: "One finds in the *Journal Etranger* of August 1755 the translation of part of this life by Ranft. One will see that it is the same thing as the first part of these reports."

8. Arsenal Library, Manuscript 4775: Table of Events of the War for the Succession of the Hereditary Lands of the Emperor Charles VI, by Sieur Brunet, 1750.

This volume contains, in a very precise and detailed form, the history of the events that occurred, during all og the War of the Austrian Succession, in Silesia, Germany, Italy, Lowlands, Scotland and during the preliminaries of Aix-la-Chapelle.

One finds in it the details of the composition and the force of the armies, as well as the organization of the command in each of them. As the author was employed in the Bureau of War, his work is not without a certain interest.

To the text are joined some manuscripts and printed documents, such as letters, orders of battle, sketches and maps among which the most important are the detailed account of the action at Melle and the surprise of Ghent sent to Sieur Brunet by an officer who took part in those two actions.

9. Communal Library of Nancy, Manuscript 637: Wars of Bohemia and Flanders (1741-1746).

The author of this work has remained anonymous. The part that it reports on the campaign of 1745 in Flanders begins with the action at Melle with the surprise of Ghent, then goes on to a dissertation on general operations.

The reading of the first accounts causes one to believe the author when he lets one know that he was, in these actions, an eye-witness. He makes a lively narration that is full of details, including a complete description to which he has attached an explanatory sketch.

In the second part, he presents a summary of the operations in their totality. He then presents a concise criticism, sometimes severe, but always reasoned and without passion.

For these reasons, it is not without interest to compare the conclusions of this contemporary with those of the anonymous author whose reports we have already discussed (See 4. above).

10. Lille Communal Library: Godefroy Portfolio No. 70.

The collection known as the *Portefeuille Godefroy* became part of the Library of Lille in 1877. It is formed of 118 portfolios and four cartons for which there is no definitive manuscript or printed catalog. One has used for the present study Portefeuille 70, bearing the inventory number 298.

This is the journal written by Jean-Baptiste Achille Godefroy de Maillart.[727] Son of a director of the Director of Reports of Lille, under Louis XIV, Godefroy de Maillart (1697-1759) was given this charge in 1726. He received in 1744 the order to go to the Army of Flanders to learn about the archives of the lands occupied by the French in the Lowlands and to look into everything that could be of use or belonged to France. One did not know, in effect, at the time of the preliminary conquests, if, upon the peace, one would not retain some of them.

Godefroy de Maillart had written a very detailed report, for the year 1745, beginning with the first days of April and ends on 14 October. Military events do not alone occupy the chronicler. Alongside his descriptions of the battles, the sieges, and the marches of the actors who have given Godefroy all the relevant details, one finds curious anecdotes on the life of Louis XV, in the middle of his troops, on the trips of the sovereign to Ghent, Ostend, Bruges, and upon his return to Paris at the end of August 1745. Godefroy de Maillart limits himself to a very detailed account, but it is lacking any type of personal commentary.

11. Archives of the Council of State (Archives of La Haye), Manuscript 1900.

Among the diverse documents that it contains, those that were used in the course of this study were: 1. The accounts of the battle of Fontenoy, including one addressed by the English General Ligonier to Lord Chesterfield and the other by Lieutenant General Aylva to the States General, 2. The orders of battle of the Pragmatic Army in the month of June 1745 and 26 July 1745, 3. Accounts of the battles at Saint-Amants and Grimberghen sent to the States General by Prince Charles von Waldeck (See 14.) All these documents are written in French.

12. Archives of the States General (Archives of La Haye): Secrete brieven van legerhoofden.

---

[727] Cf. Godefroy de Ménilglaie (Marquis de). *Les Savants Godefroy* (Paris, Didier,1873).

The totality of these documents forms the correspondence addressed to the States General by the Dutch and foreign generals employed by the Republic in the command of troops in its pay and its own service.

For the period between the months of May 1745 and February 1745 one finds there original letters written to "Their High Powers the Lords of the States General" by Königsegg, Prince von Waldeck, and the various officers given employment or a mission to the army and the fortresses of Flanders, Germany, and even Scotland. Some were written in French, others, less numerous, in English; but the greater part are written in Dutch. Finally, certain letters possess annexes among which figure above all the reports written as a result of the conferences held by the Allied generals or copies of the letters addressed by Maruice de Saxe to various officers of the Pragmatic Army and transmitted by these latter to the States General.

13. Archive of the Council of State (Archives of La Haye), Manuscript 099: Journal of the Count Schlippenbach.

The Count von Schlippenbach figured, in 1745, among the brigadiers of the Pragmatic Army operating in Flanders. He commanded, in addition, during all this campaign, a regiment of dragoons in the service of Holland.

This account is written in French. It covers the years 1743, 1744, 1745, 1746, and 1747 and it is generally lacking in detail. For the period of 1745, in particular, it relates little, with some developments, than the events that passed from the preparations made in La Haye to enter into the campaign up to 10 July. The events after this are reported only in summary.

The first part is curious. The author recounts here the numerous intrigues provoked in La Haye by the designation of the commander of the troops of the Republic. Schlippenbach shows no goodwill for the most part of the Dutch generals. He appreciates, nonetheless, the role played by these latter at Fontenoy or the siege of Tournai.

It is easy to report that the accounts of this author are, in general, exact. However, one can easily agree that Schlippenbach poorly resisted the desire overstate role he played in some circumstances that were otherwise of little importance. He also liked to present himself as frequently foreseeing events. An examination of the facts shows, however, that to the contrary, most of the time the events certainly contributed to establishing the claimed inspiration of von Schlippenbach.

With these reservations, the author remains a sober and impartial contributor who played an active, yet secondary role.

14. Archives of the Council of State (Archive of La Haye), Manuscript 1899: Journal of Prince Charles von Waldeck.

Charles Prince von Waldeck was elected, in 1745, by the States General, to the post of Commander-in-Chief of the Auxiliary Troops furnished by Holland to the Coalition. He left journals of the campaigns of 1745 and 1745, which he made in the Lowlands in this capacity. These manuscripts are written in French. The first begins on 14 April and ends on 17 November. The second opens on 31 January and ends on 2 November.

One finds in these important works a daily and very detailed account of events. The author attached all the documents that permit a deep appreciation of the role played by the Pragmatic Army in the course of these campaigns: orders of battle and of march, the states of troops, garrisons, magazines, lists of losses, distribution into winter quarters, reports of conferences held between the Allied generals, etc., etc.

The situation of Prince Charles, during the campaign of 1745, was difficult. Without being in a declared war with France, the Dutch Republic furnished the major part of the troops of the Pragmatic Army in Flanders. In the preparation of his *Journal*, the Prince expresses his constant concern about the delicate nature of his role as the principal preoccupation in reconciling his duties towards the Allies with his responsibility with regards to the Republic. Despite his bad luck, Prince Charles writes in his *Journal* with a tireless serenity. No recrimination, no criticism comes from his pen, even in the most difficult circumstances. The account contains, one can almost with say monotony, all the defeats, the capitulations, the retreats, and the deceptions of every nature added one upon the other. Prince Charles omits none of this sad series, but never does he abandon his composure. The *Journal* is a cold chronicle, methodical and impartial, which one can consult with confidence.

Otherwise, it is completed by the letters of Prince Charles contained in the *Secrete brieven van legerhoofden* (see 12) and by the documents inserted in Manuscript 1900 (See 11).

II. Printed Works

A.) Contemporary memoirs and correspondence.

1° *Mémoires du duc de Luynes sur la cour de Louis XV (1725-1758)*, published L. Dussieux et Eudore Soulié. — Paris, 1860-1865, Didot, 17 vol. in-8°.

2° *Journal et Mémoires du marquis d'Argenson*, published for the Société de l'histoire de France, par E.-S.-B. Ratheky. — Paris, 1859-1867, Renouard, 9 vol. in-8°.

3° *Chronique de la Régence et du Règne de Louis XV (1718-1763) ou Journal de Barbier, avocat au Parlement de Paris.* — Paris, 1857, Charpentier, 8 vol. in-8°.

4° *Les Rêveries ou Mémoires sur l'Art de la guerre de Maurice de Saxe, dédiées à MM. les officiers généraux*, by M. de Bonneville, Captain of H.M. the King of Prussia — La Haye, 1756, Pierre Gosse Junior, 1 vol. in-folio.

5° *Lettres et Mémoires choisis parmi les papiers originaux du maréchal de Saxe.* — Paris, 1791, Smils et Cᶜ, 5 vol. in-8°.

These letters and memories were published by Grimoard and come, for the most part, from the Archives of War. One can agree that there are great inaccuracies in the reproduction and numerous omissions.

B.) Documents.

1° *Ordonnances du Roy portant règlement pour l'habillement, équipement et armement de la cavalerie avec le projet d'instruction pour les évolutions de la cavalerie et celui d'instruction.* — Metz, 1733, Brice Antoine, 1 vol. in-16.

2° *Règlement provisionnel pour le service de la cavalerie en campagne.* — Paris, 1744, Imprimerie Royale, 1 vol. in-12.

3° *Règlement provisionnel pour le service de l'infanterie en campagne.* — Paris, 1744, Imprimerie Royale, 1 vol. in-4.

4° Bibliothèque du Ministère de la Guerre : *Collection des ordonnances Royales.*

In addition, the ordnances come from the Royal authority,

the volumes of this collection include numerous copies, made by the
Marquis de Saujon, decrees or organic circulars figuring in the Cangé
Collection of the Bibliothèque nationale.

C.) Works subsequent to the campaign of 1745.

1° Z***, Chevau-léger de l'une des compagnies d'ordonnances de la Gendarmerie : *La Conquest des Pays-Bas par le Roy
dans la Campaigne de 1745 avec la prise de Burxelles en 1746.* La
Haye, 1747, 1 vol. in-16.
By Zambault, contains a map of Fontenoy and the nominal
states of the composition and command of the army.

2°Baron d'Espagnac, Gouverneur de l'Hôtel
Royal des Invalides: *Histoire de Maurice de Saxe, Duc de Courlande
et de Sémigalle.* Paris, 1775, P.D., Pierres, 3 vol. in-4°.
The author of this work, whose documentation is defective,
filled, during the 1745 campaign, the functions of aide maréchal de logis
to the staff of Maurice de Saxe.

3°Alfred Ritter von Arneth: *Geschichte Maria Theresia.* – Vienna, 1863-1879, Braumüller, 10 vol.
in-8°.
Included in the appendices are numerous extracts from official
dispatches drawn from the Vienna Archives.

4°Count C.F. Witzbum d'Eckstardt: *Mau-
rice de Saxe et Marie-Joseph de Saxe dauphine de
France.* – *Lettres et documents in*édits *des Archives de
Dresde.* – Leipzig, 1867, L. Denicke, vol. in-8°.

5°Duc de Broglie: *Marie-Th*érèse *impératrice
(1744-1746).* – 3rd edition. Paris, 1893, Calmann-Lévy, 2 vol. in-12°.

6° Duc de Broglie: *Maurice de Saxe et le marquis d'Argenson.*
– Paris, 1893, Calmann-Lévy, 2 vol. in-12°.

7° *Oesterreichischer Erbfolg-Krieg 1740-
1748,* herausgegeben von der Direction des K. und
K. Kriegsarchiv. – Vienna, 1896, Seidel und Sohn,
8 vol. parus in-8°.

8° P. Colin, Brevet Captain of Artillery, *Les
Campagnes du mar*échal *de Saxe.* – Paris, 1801-1906, Chapelot,
3 vol. in-8°.
Abundant documentation drawn from the War Archives, those

of La Haye, Lille, Nancy, etc.

Abbreviations Employed for the Source Manuscripts.

Historical Archives of the Ministry of War	A.H.G
Archives of La Haye...	A.L.H
Volumes, 3084, 3085, 3086, 3087, 3088, 3089, 3090, 3091, 3092, 3135,	
17 and 18 (see 1, above)	Corresp. A.F.
Volumes 309, 3099, 3100 (see 1 above)	Corresp. A.A.
Volume 3093 (see 3, above)	De Vault
Memoires written in December by an anonymous author (See 4, above)..	R.A
Brézé's Journal (See 5, above)	Brézé.
Pon's Journal (See 6, above)	Pons.
Wars of Bohemia and Flanders, by an anonymous author (See 9, above)...	R.A'.
Godefroy's Portfolio (See 10, above)	Godefroy
Secretbrieven van legerhoofden (See 12, above)	Corresp. G.A.
Count von Schlippenbach's Journal (See 13, above)	Schlippenbach
Prince Charles von Waldeck's Journal (See 14, above)	Waldeck.

A general map of the theater of operations was prepared from the map of France, 1/345,600, by Capitaine, reduced to 1/86,400 by Cassini and Ferraris, prepared around 1795. In the course of the work one has updated the names of the localities with those that Capitaine employed.

Other maps bearing the indication of their source, except for the plans of the attacks at Dendermonde and Brussels which were drawn from the work of Baron d'Espagnac (see: imprinted works, c. 2.) and the sketches of the action at Mellen were prepared according to a map of the terrain found in the archives of maps at the Ministry of War.

INDEX